Insulation Co-ordination in
High-voltage Electric
Power Systems

Insulation Co-ordination in High-voltage Electric Power Systems

W. DIESENDORF Dr. Techn. Sc.

Senior Lecturer,
School of Electrical Engineering,
University of Sydney,
formerly
System Design Engineer,
Snowy Mountains Hydro-electric
Authority, Australia

LONDON

BUTTERWORTHS

THE BUTTERWORTH GROUP

ENGLAND
Butterworth & Co (Publishers) Ltd
London: 88 Kingsway, WC2B 6AB

AUSTRALIA
Butterworths Pty Ltd
Sydney: 586 Pacific Highway, NSW 2067
Melbourne: 343 Little Collins Street, 3000
Brisbane: 240 Queen Street, 4000

CANADA
Butterworth & Co (Canada) Ltd
Toronto: 14 Curity Avenue, 374

NEW ZEALAND
Butterworths of New Zealand Ltd
Wellington: 26-28 Waring Taylor Street, 1

SOUTH AFRICA
Butterworth & Co (South Africa) (Pty) Ltd
Durban: 152-154 Gale Street

First published in 1974

© Butterworth & Co (Publishers) Ltd., 1974

ISBN 0 408 70464 0

Printed in Hungary

Preface

This book is concerned with one important area in the larger field of high-voltage insulation: the insulation design and surge performance of high-voltage transmission lines and stations, customarily referred to as 'insulation co-ordination'. In this area one depends on a mass of observational and experimental data as a basis of analysis, and there is a continual accretion of knowledge, scattered over many engineering publications. The recent trend to ultra-high voltages and statistical methods has brought a new spate of papers. Engineers responsible for the planning and operation of electric power systems, excepting the specialists of the largest undertakings, cannot as a rule afford the time to delve into this vast amount of literature. The author's purpose is to provide a vade-mecum to help them in their work and assist their understanding of the essentials. This objective, he hopes, will also be welcomed by utility engineers generally, students of power system engineering and equipment design engineers who wish to inform themselves of the wider electrical environment in which their products will have to exist.

An introduction to overvoltages in power systems and elementary probability theory is included. References and bibliography point the way to further study.

Sydney W.D.

Acknowledgements

The seeds of this book were sown a number of years ago when the author lectured on insulation co-ordination to a refresher school on Power System Engineering initiated by Professor C. E. Moorhouse, Head of the Department of Electrical Engineering at the University of Melbourne. In repeat schools held at various Australian Universities, the section on insulation co-ordination was contributed to by Drs. M. Darveniza and T. M. Parnell, both of the University of Queensland. Building on the material so accumulated, the writer enlarged and consolidated it on the occasion of a seminar he conducted at Rensselaer Polytechnic Institute, encouraged by Dr. E. T. B. Gross, Chairman of Electric Power Engineering, and aided by stimulating discussions with Drs. M. Harry Hesse and T. S. Lauber, both professors of Electric Power Engineering at Rensselaer. These lectures were published in 1971 by Rensselaer Bookstore, Troy, N.Y., under the title 'Overvoltages on High Voltage Systems (Insulation Design of Transmission Lines and Substations)'.

The present book is the result of a complete rewrite and inclusion of most recent material. The author is indebted to all colleagues named above for their contributions which are contained in this book in some form or other, in particular to Dr. Gross for permission to re-use material. Special thanks are extended to Dr. Darveniza who co-authored Chapters 4 and 5 and provided the examples relating to the transmission line of Appendix B, drawn from his graduate course in lightning protection at the University of Queensland; also for helpful advice and criticism generally. Last but not least to my wife Margaret who used her linguistic acumen to uncover technical inconsistencies.

Permission to use other previously published material is acknowledged where it appears.

Contents

1   **Introduction**   **1**

2   **Overvoltages**   **3**

     2.1   Lightning surges   4
     2.2   Temporary overvoltages   9
     2.3   Switching overvoltages   15

3   **Disruptive Discharge and Withstand Voltages**   **27**

     3.1   Introduction   27
     3.2   Self-restoring insulation   28
     3.3   Non-self-restoring insulation   42

4   **Lightning Overvoltages on Transmission Lines**   **47**

     4.1   Introduction   47
     4.2   Strokes to nearby ground   49
     4.3   Shielding   50
     4.4   Strokes to towers   55
     4.5   Strokes to ground wires   60
     4.6   Practical methods of determining the voltages stressing line insulation   61
     4.7   Attenuation and distortion of lightning surges   63

5   **The Lightning Performance of Transmission Lines**   **70**

     5.1   Introduction   70
     5.2   The flashover rate of unshielded lines   71
     5.3   The flashover rate of shielded lines   72
     5.4   Outage rate and sustained outage rate   80

6   **The Switching Surge Design of Transmission Lines**   **84**

     6.1   General   84
     6.2   Tower insulation design   88
     6.3   Application to future ultra-high voltages   92

CONTENTS

7 The Insulation Co-ordination of High-voltage Stations   **95**

    7.1   Principles   95
    7.2   Overvoltage protective devices   96
    7.3   Stations with protected zone   104
    7.4   Stations without protected zone   115
    7.5   Cable-connected equipment   115

Bibliography   **118**

Appendix A Propagation of travelling waves   **119**

**Appendix B Data for 220 kV transmission line used in Lightning performance
    calculations**   **125**

Index   **127**

1

Introduction

The reliability of supply provided by an electric power system, as judged by the frequency and duration of supply interruptions to its customers, depends to a great extent on the surge performance of the system. Although there are many other causes of interruptions, breakdown of insulation is one of the most frequent.

If the insulation were subjected only to the normal operating voltage which varies within quite narrow limits, there would be no problem. In reality, the insulation has to withstand a variety of overvoltages with a large range of shapes, magnitudes and durations. These various parameters of overvoltages affect the ability of insulation to withstand them. The problem before us is therefore:

to ascertain the magnitudes, shapes, frequency and duration of overvoltages, and the changes they undergo when travelling from the point of origin to the equipment affected;

to determine the voltage withstand characteristics, in respect to these overvoltages, of various types of insulation in use;

to adapt the insulation strength to the stresses.

It is soon found that it is not always possible to provide insulation which will withstand the highest stresses that may occur. An economic limit intervenes well before the technical limit. The engineer places this limit at the point at which the cost of achieving a further improvement in reliability cannot be justified by the savings the reduced number of breakdowns may bring, which are at any rate difficult to assess in money terms. He deliberately accepts a certain probability of breakdown in the design of power systems. In this respect his design philosophy differs from that of the civil engineer who designs those structures whose failure could have catastrophic consequences to withstand all foreseeable stresses.

On the other hand, the design of a power system should be such that, when breakdowns are inevitable, they are confined to locations where they cause minimum damage and the least disturbance to operation.

The worst damage is caused by breakdown of solid insulation in a confined space, the least is suffered by insulation that can recuperate, and in this case not every breakdown leads to a circuit breaker operation.

Disturbances to supply after insulation failure can be minimised by such measures as fast and sensitive relay protection, provision of duplicate or alternate supply and automatic reclosing, but these techniques are outside the scope of this book.

2

Overvoltages

Overvoltages can be impressed upon a power system by atmospheric discharges, in which case they are called 'lightning overvoltages', or they can be generated within the system by the connection or disconnection of circuit elements or the initiation or interruption of faults. The latter type are classified as 'temporary overvoltages' if they are of power or harmonic frequency and sustained or weakly damped, or as 'switching overvoltages' if they are highly damped and of short duration. Because of their common origin, temporary overvoltages and switching surges occur together, and their combined effect is relevant to insulation design. The probability of a coincidence of lightning and switching surges, on the other hand, is small, and can be neglected.

The prospective magnitudes of lightning surges appearing on transmission lines are not much affected by line design; hence lightning performance tends to improve with increasing insulation level, i.e. system voltage. The magnitudes of switching surges, on the other hand, are substantially proportional to operating voltage. As a consequence, there is a system voltage at which the emphasis changes from lightning to switching surge design; this point is reached at approximately 300 kV. In the 'extra-high voltage' range, up to the highest existing system voltage of 765 kV, both lightning and switching overvoltages have to be considered. For the 'ultra-high voltages' at present being investigated, switching surges will be the main, though not the sole criterion; insulator pollution in particular will remain an essential design factor.

The importance of switching surges is accentuated by the fact that the switching impulse strength of external insulation and airgaps, from about 1 MV up, no longer increases in direct proportion to striking distance, unlike lightning impulse strength which does.

For the study of overvoltages, at least a basic knowledge of the laws of surge propagation is desirable. A number of excellent textbooks[1,2] exist (also Chapter 15 of [7]); Appendix A is therefore limited to a summary of the principal relations.

2.1 LIGHTNING SURGES

2.1.1 Lightning phenomena

Before considering the effects of lightning discharges on power systems it is necessary to inquire into their natural characteristics[3-7].

During thunderstorms, positive and negative charges become separated by the interplay of air currents, ice crystals in the upper part of the cloud and rain in the lower part. This process is the subject of many theories but of interest for our purposes are the following observable facts: the great mass of cloud becomes negatively charged, with a layer of positive charge at the top, which is typically 9–12 km high, and a small positive inclusion near the base. The negative charge centres may be from 500 to 10 000 m above the ground surface, the cloud base may be as low as 150 m. The potential of thunderclouds has been estimated to be at least 100 million volts.

A lightning stroke to earth usually appears to the eye as a single luminous discharge, although sometimes rapid fluctuations of light intensity can be seen. Photographs with rotating-lens cameras reveal that most strokes are followed by repeat strokes which travel along the path established by the first discharge, at intervals of 0.5–500 ms.

The first component of a stroke is initiated by a 'stepped leader' which starts in a cloud region, where a local charge concentration causes the voltage gradient to reach the critical breakdown value. In dry air at sea level the breakdown gradient is 30 kV/cm but in a region filled with water droplets, at the higher altitude, it is approximately one third of this value. The leader consists of a highly ionised core or channel. It is preceded and surrounded by a corona envelope which has a diameter of about 30 m and extends about 50 m in front of the channel. The leader tip moves rapidly in steps of about 50 m and pauses after each step for a few microseconds while streamers emanating from it charge the corona sheath that enables it to proceed. The average speed of propagation is about 150 km/s. If the distance from stroke centre to ground is say 3000 m, it may take the leader 20 ms to bridge it. The steps are straight but, by the action of space charges and wind, each new step changes direction. Branches appear which may terminate in midair whilst the main channel continues a zigzag path to earth. During its relatively slow descent the stepped leader deposits a negative charge along its path. As the leader head approaches the ground, the positive charges induced in the general target area intensify; however, the point of impact remains indeterminate until the leader has arrived at a certain striking distance from the ground surface. At this distance, the charge in the leader produces at the ground 'electrode' a gradient sufficient to cause breakdown. Because of the non-uniformity of the field, the critical gradient, averaged over the striking distance, is of the order of one sixth the breakdown gradient in dry air.

At this stage short positive streamers rise from the earth, preferring high projections as their starting points. When the negative leader encounters the positive streamer, an intensely luminous discharge starts from earth to cloud, travelling at a velocity varying from 10 to 50% of that of light, depending on the charge density. This discharge is referred to as the 'return stroke'. What happens is that the negative charge laid down by the leader stroke along its path is being rapidly neutralised by the upward-moving positive charge previously induced in the ground and in objects on the ground. The luminous point indicates the boundary to which the positive charge has penetrated at any instant. The current at the point of impact can be considered either as a negative current flowing into the ground or a positive current flowing out of it. For a small proportion of lightning flashes the downward current is positive. The current in the return stroke is in the range from a few kiloamperes to about 260 kA. It is very much higher than the current in the leader; since the same quantity of charge is involved in both phenomena, the currents are in the ratio of their velocities.

After the first return stroke has completed the discharge of portion of the cloud, streamers develop from negatively charged cloud regions at higher altitude, and create channels by which they are connected to the still ionised and heated path of the first stroke. A 'dart' leader develops between cloud and earth which follows this path without branching, and at a higher velocity than that of the stepped leader. Upon striking the ground, a second return stroke travels back to the cloud. This process may repeat itself many times. From 30 to 80% of strokes have at least two component flashes, about 20% have three to five components but there can be up to 40.

It has been established that many more discharges take place intra-cloud and from cloud to cloud than between cloud and ground. The ratio of cloud flashes to groundflashes varies between 1.5–3 in temperate zones and 3–6 in tropical climates[8].

2.1.2 Lightning stroke characteristics

Starting in the late 1920s a great number of field investigations were conducted to ascertain the characteristics of lightning which affect overhead transmission line performance (see Chapter 16 in [7]). Unfortunately even today the available information is neither definite nor complete.

Lightning current oscillograms indicate an initial high-current portion, which is characterised by short front times of up to 10 μs; the front is usually concave upwards. The high-current portion lasting some tens of microseconds is followed by a long-duration, low-current tail which may last hundreds of milliseconds and is responsible for thermal damage, so called 'hot' lightning. A typical oscillogram is shown in *Figure 2.1*. Lightning currents are measured either directly on high towers or buildings, which are not really typical of transmission lines, or on the four corner legs of transmission towers, which is inac-

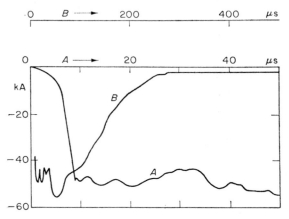

Figure 2.1. Typical lightning current oscillogram (after Berger[6])

curate because of the unequal division of leg currents and the presence of ground wires and adjacent towers. Based on the results of many investigations an AIEE Committee[9] has produced the frequency distribution of current magnitudes, Curve 1 of *Figure 2.2*, which is widely accepted for performance calculations. More pessimistic curves have been proposed, such as that by Anderson (Chapter 8 of [10]), shown as Curve 2 on *Figure 2.2*. It is interesting to observe that recent data (Curve 3) compiled by a Study-committee of the International Conference on Large High-Tension Electric Systems (CIGRE), based largely on strokes to tall objects, suggest that the probability of large stroke currents (above 100 kA) is much greater than indicated by either curve. It can be shown theoretically that tall objects attract a larger proportion of high-current strokes and this would explain the shift of the frequency distribution curve towards higher currents[11].

Another important characteristic is the time to crest of the current waveshape. *Figure 2.3* reproduces probability distribution curves from two sources which are reasonably consistent. There is evidence that very high stroke currents do not coincide with very short times to crest. Field data[7] indicate that 50% of stroke currents have a rate of rise exceeding 7.5 kA/μs and 10% exceed 25 kA/μs. The mean duration of stroke currents above half value is 30 µs and 18% have longer half times than 50 µs.

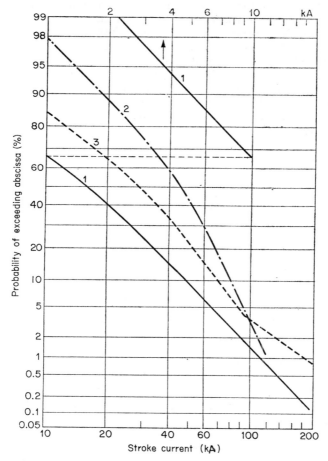

Figure 2.2. Cumulative distributions of lightning stroke current magnitudes:
1, (after AIEE committee[9]),
2, (after Anderson[10]),
3, (after Popolansky, Electra (CIGRE), No. 22, 139-147 (1972))

The risk of lightning strikes to electrical installations is necessarily related to the degree of thunderstorm activity. The only indicator readily available from national meteorological services throughout the world and the World Meteorological Organisation in Geneva[12] is isokeraunic level or thunderdays (TD), defined as the number of days in a year, or month, when thunder is heard at any particular location.

Weaknesses of this measure, from the transmission engineer's point of view, are that it does not distinguish between ground strokes and cloud strokes and does not recognise the varying intensities and durations of thunderstorms. A somewhat better measure is thunderhours, but the most appropriate measure is groundflash density (N_g). Attempts are being made to obtain sufficient statistical data of N_g and for this purpose lightning flash counters have been developed[8, 13]. As these respond to nearby cloudflashes as well as groundflashes, they have to be calibrated for groundflashes in each region, by optical or other observations[8, 14]. Until more data from this or similar devices become available, estimates of lightning intensity will continue to be based on thunderday level.

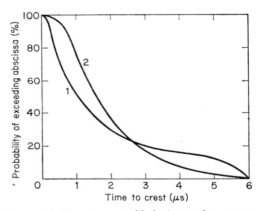

Figure 2.3. Time to crest of lightning stroke currents:
1, (after McEachron, AIEE Trans. 60, 885 (1941)),
2, (after Anderson[10], p. 288)

A frequently employed but imprecise empirical formula for groundflash density is

$$N_g = (0.1-0.2)\,(\text{TD}) \text{ strokes/km}^2 \text{ year} \qquad (2.1)$$

More reliable correlation has to await the results of continuing field recordings.

In Great Britain, Europe and the Pacific Coast of North America, TD is in the range 5–15; N_g is of the order of 1.0–2.0. In the USA the thunderday level increases towards the central and eastern States, where levels from 30 to 50 are prevalent, and reaches a peak of 80 in Florida. Areas of even higher thunderday levels exist in South Africa and South America.

The interaction between lightning strokes and the transmission system will be discussed in Chapters 4 and 5.

2.2 TEMPORARY OVERVOLTAGES

2.2.1 Introduction

The significance of temporary overvoltages in respect to insulation co-ordination lies in the requirement that surge diverters (lightning arresters) must be able to reseal against sustained voltages or risk destruction. Since the protective level of any particular kind of surge diverter is proportional to the reseal voltage, the insulation level and cost of equipment depends indirectly on the magnitudes of temporary overvoltages (see 7.3.2). In the extra high voltage range, temporary overvoltages cum switching surges determine the insulation of transmission lines and consequently their dimensions and cost.

The main causes of power frequency overvoltages of interest in the present context are:
sudden loss of load; disconnection of inductive loads or connection of capacitive loads; Ferranti effect, and unbalanced ground faults.

Overvoltages with frequencies not very different from power frequency arise when shunt-compensated transmission lines are disconnected and trapped charges oscillate between line capacitance and reactor inductance. Higher harmonic-frequency oscillations can be excited by the magnetising currents of unloaded transformers, and sub-harmonic oscillations may be caused by series capacitors resonating with lightly loaded transformers or shunt reactors. Single-phase switching and broken conductors can cause ferro-resonance overvoltages but in this and other rare cases prevention should be the aim to avoid excessive expenditure on insulation.

The duration of temporary overvoltages may vary from a few cycles, if intertripping or voltage-dependent relay protection is provided, or a few seconds, if reduction depends on automatic voltage regulators, to much longer periods if human intervention is relied upon.

Severe power-frequency overvoltages can be caused by the combination of full-load rejection with a change from an inductive to a capacitive load (e.g. by opening of a circuit-breaker at the far end of a loaded transmission line), and a simultaneous ground fault. If other lines or a local load offer some outlet for power, complete rejection of the power station load is improbable. Generally speaking, the more a system is 'meshed', the lower are the overvoltages caused by load rejection. This situation prevails in well-developed systems but is unlikely in the initial stages of a new higher-voltage system. On long transmission lines the position is greatly eased by shunt reactors, either permanently connected or rapidly connectable in case of overvoltage by means of special sparkgaps. The expenditure on reactors can very often be justified by the need to reduce charging MVAr (see 2.2.4).

2.2.2 Load rejection

When a transmission line carrying a substantial portion of the output of a power station is switched off, the generators will speed up and the voltage will rise. The speed governors and the automatic voltage regulators will intervene in the sense of restoring normal conditions. An exact determination of the maximum bus voltage reached requires a full knowledge of the parameters of the machines, governors and excitation systems, as well as mathematical aids such as digital or analogue computers. Relevant studies are described in[15-17].

For an approximate evaluation, which may suffice in a first approach, it can be assumed that initially the voltage behind subtransient reactance remains unchanged at its value before the incident but after a few cycles the voltage E_d' behind transient reactance becomes the constant driving voltage. Neglecting losses and the brief subtransient period, the relationship between E_d' and the sending end bus voltage (V_1) becomes

$$V_1 = (f/f_0)E_d'/[1 - (f/f_0) X_s/\overline{X}_c] \qquad (2.2)$$

This equation accounts for the voltage rises associated with overspeeding through its effect on E_d' and the variation of inductive and capacitive reactances with frequency. X_s is the reactance between E_d' and V_1 (usually the sum of generator transient reactance and transformer reactance), $\overline{X}_c$ is the capacitive input reactance of the open-circuited line at the increased frequency, and f/f_0 the ratio of instantaneous frequency at the time maximum voltage is reached to rated frequency. The difficulty is in determining f.

For turbo-generators the maximum speed rise on full-load rejection is approximately 10% and it occurs in less than a second. In this time the rotor flux has not changed much and the maximum voltage can be approximately calculated from equation 2.2, using the instantaneous frequency corresponding to the maximum speed rise for the load rejected.

In the case of waterwheel generators, the maximum speed rise on full-load rejection can be as high as 60% but it may take up to 10 s to reach it. A fast acting voltage regulating system will have begun reducing excitation well before then and the maximum voltage may occur perhaps one second after rejection. Failing a more accurate method, the instantaneous frequency (f in equation 2.2) may be estimated by assuming a linear increase until maximum speed is reached.

The voltage at the sending end is greatly affected by (a) line length, (b) short-circuit MVA (S_{sc}) at the sending end bus (which is determined by X_s) and (c) the reactive generation of the line (determined by X_c and any series or shunt compensation). These factors are illustrated in *Figure 2.4*. The beneficial effect of shunt reactors is clearly shown.

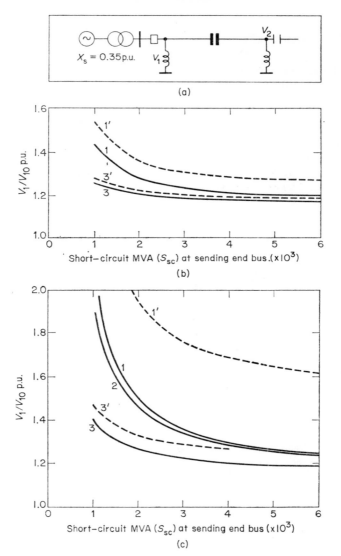

Figure 2.4. Power frequency voltages on load rejection—(a) diagram, (b) 300 km line, (c) 600 km line:

1, 1', no compensation;
2, 50% series capacitor compensation;
3, 3', 50% series capacitor and 70% shunt reactor compensation;
1, 2, 3, sending end; 1', 3', receiving end;
initial voltages $V_{20} = V_{10} = 400$ kV = 1 p.u.;
initial load = $0.33S_{sc}$; 10% speed rise; losses neglected

2.2.3 Ferranti effect

The Ferranti effect of an uncompensated line is calculated by the approximate formula:

$$V_2/V_1 = 1/\cos\beta l \qquad (2.3)$$

where V_2 is the voltage at the open receiving end and β the phase constant; the latter is 6°/100 km at 50 Hz, 7.2°/100 km at 60 Hz. For compensated lines the calculation is somewhat more complex. The mitigating effect of compensation, particularly by shunt reactors, is seen in *Figure 2.5*. The combined influence of load rejection and

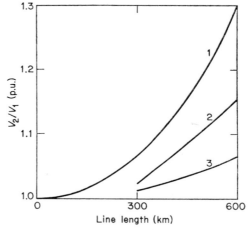

Figure 2.5. Ferranti effect (see Figure 2.4a for diagram):
1, no compensation,
2, 50% series capacitor compensation,
3, 50% series capacitor and 70% shunt reactor compensation

Ferranti effect on receiving end voltage is shown by the broken lines in *Figure 2.4*.

2.2.4 Self-excitation

A potentially dangerous condition exists when the reactive power generated by the unloaded transmission line, adjusted for speed rise, exceeds the ability of the generators on line to absorb it. Under capacitive loading the armature reaction assists the field; to keep the steady-state bus voltage to its nominal value, a very low or even negative field excitation may be needed. Certain criteria have to be met to maintain voltage stability under these circumstances. For uncompensated lines:

(1) If $X_c(f_0/f)^2$ is greater than X_d (the direct-axis synchronous reactance including step-up transformer reactance), the excitation is positive and can be controlled stably even by manual means.

(2) If
$$X_q < X_c (f_0/f)^2 < X_d \qquad (2.4)$$

(where X_q is the quadrature-axis synchronous reactance including transformer reactance), it is a case of 'slow self-excitation' and the voltage can be controlled stably with a continuously acting automatic voltage regulator.

(3) If
$$X_c (f_0/f)^2 < X_q \qquad (2.5)$$

the voltage builds up rapidly and uncontrollably, a case of 'fast self-excitation'; it is limited only by generator and transformer saturation[18, 19]. This condition must be prevented at all costs, either by ensuring that sufficient generator capacity is on the bus, or by compensating part of the charging MVAr of the transmission system with shunt reactors. The maximum frequency reached is to be used in the above equations.

2.2.5 Ground faults

A single line-to-ground fault causes a rise in the voltages to ground of the healthy phases which depends mainly on the effectiveness of neutral earthing[20]. For isolated neutral or arc suppression coil systems, the potentials of the healthy phases can exceed the line-to-line voltage; for solidly grounded systems they will increase above their normal values but remain below line-to-line voltage. Double line-to-ground faults may also produce increases in line-to-ground voltages.

A measure of the voltage rise caused by single line-to-ground faults is the 'earth fault factor', defined as the ratio of the higher of the two sound-phase voltages to the line-to-neutral voltage at the same point of the system, with the fault removed[21]. It is calculated by symmetrical component methods, or obtained from graphs such as *Figure 2.6* which is taken from Chapter 18 of[7]; this reference contains a set of such graphs for various resistance/reactance ratios.

An 'effectively earthed' system is characterised in international and many national standards by an earth fault factor not exceeding 1.4. Solidly earthed systems normally comply with this rule. From the curves in *Figure 2.6*, a system is effectively earthed if

$$X_0/X_1 \leqslant 3.0 \quad \text{and} \quad R_0/X_1 \leqslant 1.0 \qquad (2.6)$$

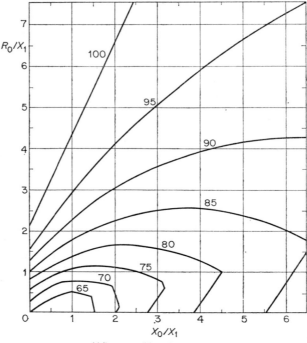

Voltage conditions for $R_1 = R_2 = 0.1 X_1$

Figure 2.6. Maximum line-to-ground voltage at any fault location and under any fault condition of grounded-neutral system. Numbers on curves are maximum line-to-ground voltage of any phase, in per cent of line-to-line voltage. Effect of fault resistance was taken into account at the value which gives the maximum voltage to ground on any phase:

R_0, zero-sequence resistance,
X_0, X_1, X_2, zero-sequence, positive-sequence and negative-sequence reactances, $(X_1 = X_2)$ (courtesy Westinghouse Electric Corp.[7], p. 626)

2.2.6 Saturation effects

Transformer magnetising currents increase rapidly for voltages above rated; they can equal rated current at 1.5 rated voltage. Overvoltages caused by load rejection may be reduced if the receiving-end transformer remains connected, i.e. if the low-voltage circuit-breaker opens. The problem is amenable to a graphical solution provided the saturation characteristic is known.

The effect of transformer magnetising currents is not always favourable. Saturated transformers inject substantial harmonic currents into the system. Typically, at 2 Wb/m², corresponding to say 1.2 p.u. voltage, the third harmonic may be 65%, the fifth 35% and the seventh 25% of the exciting current. These harmonic currents are impressed on a network which, seen from the transformer magnetising branch, contains inductance in the power station branch and capacitance in the transmission line branch. For harmonics that are multiples of three, the zero-sequence values are effective; delta-connected windings help to suppress them. Generators are represented by the subtransient reactance, adjusted for frequency. If the parallel resonant frequencies of the system lie near to a harmonic frequency, large harmonic voltages appear in the network which add to the fundamental-frequency overvoltages, and may outweigh the reduction in the fundamental due to saturation. For the higher harmonics, series resonance between the transformer magnetising inductance and the line capacitance may occur which can produce even higher overvoltages. Because of the nonlinearities, the analysis is best performed by transient network analyser. A typical case is described in[23].

2.3 SWITCHING OVERVOLTAGES

2.3.1 Introduction

It has already been pointed out that switching overvoltages are the criterion by which the insulation of extra high voltage (e.h.v.) systems has to be designed. The reduction of switching surges is therefore an economic necessity.

In the past, circuit-breaker design was directed towards reducing the overvoltages caused by the interruption process. As these efforts were successful, it was found that surges arising on energising e.h.v. transmission lines became more critical and circuit-breakers were developed to control these closing surges. Indications are that in the future, overvoltages accompanying the initiation of short-circuits, which are uncontrollable, may establish the next lower limit[24, 25]. The continuing reduction in switching surge magnitudes may result in lightning performance again increasing in relative importance. The absolute lower limit, as far as insulation exposed to the atmosphere is concerned, will probably be set by insulator pollution.

The peak magnitude of a phase-to-ground switching overvoltage can be expressed in 'per unit' relating it to the peak voltage to ground. A phase-to-phase overvoltage is also expressed in terms of the highest voltage peak to ground. Quite often the term 'overvoltage factor' is used to indicate the ratio of the overvoltage to the peak of the system voltage prior to or after the transient. This voltage may of course differ considerably from the highest voltage for equipment and to avoid misunderstandings the reference voltage and the conditions of the case ought to be clearly stated.

2.3.2   Characteristics of switching overvoltages

Switching surges are of a great variety of shape, magnitude and
duration, corresponding to the great variety of initiating events. For
a particular event, their parameters are determined by both the system
and the characteristics of the switching device. The shape can be uni-

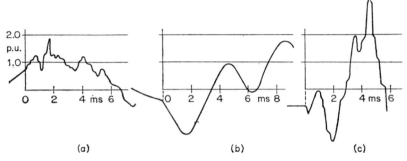

Figure 2.7. Typical switching surge waveshapes:
(a) fault initiation,
(b) fault clearing,
(c) line energising

polar, oscillatory or quite irregular and they may be superposed on
power frequency or temporary voltages. *Figures 2.7, 2.10* and *2.11*
give some examples of waveshapes.

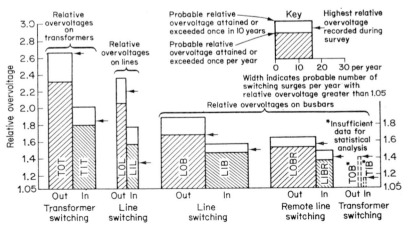

*Figure 2.8. Statistical data on switching surges in the British 132 kV grid (courtesy
White, Reece[26])*

The magnitude of a switching overvoltage for a given operation depends on the point on the voltage or current wave when switching occurs. Conditions producing the highest peaks are relatively rare; therefore, to gain an appreciation of the range of voltage magnitudes, they are represented by frequency distribution curves. Much statistical information has been collected of which *Figures 2.8* and *2.9* are typical.

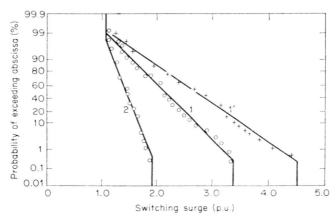

Figure 2.9. Typical switching surge distribution curves:
1, 1', uncontrolled surges,
2, controlled surges

AIEE and IEEE Committee reports[27, 28, 29] give extensive surveys and condensations of available information on switching surge characteristics, obtained by field records and transient network analyser (TNA) studies.

The collection of field records in normal operation is a long-term process, while staged tests are costly to arrange and disturb operation; in any case, these methods yield only small statistical samples. One is therefore thrown onto computer or transient network analyser methods which, by a systematic variation of relevant parameters, can produce a complete statistical distribution, including the maximum values which are less likely to be obtained in the field. Field tests are, however, extremely valuable for verification of the input data and of the mathematical model or simulation used, and have been staged for this purpose. The literature contains many examples of these procedures[29].

Some of the highest switching surges are recorded when interrupting inductive currents before their natural zero; this 'chopping' is most likely to occur for small currents, such as shunt reactor and transformer magnetising currents. The energy involved is small enough to be readily absorbed by surge diverters normally connected near the terminals of important transformers. The use of resistance-switching also assists in rendering these overvoltages harmless.

The interruption of capacitive currents, be they load currents of capacitor banks or charging currents of long unloaded transmission lines, can lead to a high voltage build-up if circuit-breaker restrikes occur at unfavourable points of the voltage wave. Modern circuit-breakers have been made practically restrike-free; thereby this cause of serious overvoltages has also been eliminated.

The switching operations of greatest concern in e.h.v. systems can be classified as follows:

(1) Line energisation, with line open-circuited at the far end or terminated in an unloaded transformer.

(2) Line re-energisation, with trapped charge on line from previous interruption.

(3) Load rejection, by circuit-breaker opening at far end, possibly followed by disconnection at sending end.

(4) Transformer switching at no-load, or with secondary load of shunt reactors; also high-voltage reactor switching.

With modern breakers and correct system design, only cases (1) and (2) are critical; they are discussed in some detail in 2.3.3.

2.3.3 Generation of energising transients

The magnitude and waveshape of energising transients depend on many factors: the length of line switched, the electrical characteristics of the source and the system configuration generally, the presence of trapped charges and, of course, any measures taken to control them.

Single-phase line, infinite bus. The simplest case, energisation of a single-phase line from an infinite bus, shall serve to explain the mechanism. Assuming that contact is made when the source voltage is V_g cos ϕ, a sine wave starting with a step of this value travels down the line and is doubled at the open end. The next reflection at the source end has a reflection factor of -1. The voltage at the receiving end is therefore composed of a succession of superimposed sine waves, with step fronts of $2V_g$ cos ϕ, alternating in polarity, and each delayed in respect of the preceding wave by an angle $2\beta l$ corresponding to twice the line travel time $2T$. The angle β is the phase factor of 2.2.3. *Figure 2.10* shows in solid lines the resulting waveshapes for $\phi = 0$. An instructive graphical treatment has been given by Lauber[30, 31]. He proved that the maximum possible receiving end voltage is $2V_g/\cos \beta l$; this expression will be recognised as twice the steady-state Ferranti voltage peak, equation 2.3. Attenuation will, however, cancel the Ferranti effect for all but the longest lines. If the switch is closed when the instantaneous supply voltage is small the vertical steps seen in *Figure 2.10* are proportionately reduced.

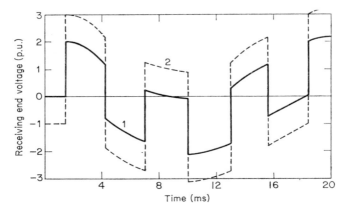

Figure 2.10. Receiving end voltage when energising a 430 km line from an infinite source (after Bickford and Doepel[32]):
1, without trapped charge,
2, with trapped charge

Trapped charges. When an open-circuited line is switched out without a restrike occurring, interruption will take place at current zero, with the supply voltage at peak value (V_g). Owing to the Ferranti effect, the disconnected line may assume an even higher voltage ($V_t > V_g$). The trapped charge decays slowly if there is no other discharge path than leakage over the insulators; in dry weather it may take minutes for the charge to dissipate. It is therefore quite possible that at the moment of line re-energisation after a short delay, as in the case of high-speed reclosing, V_g and V_t are of opposite polarities and near their peaks. Assume $V_g = 1.0$ p.u. and $V_t = -1.0$ p.u. The voltage difference of 2.0 p.u. will cause a wave with this initial step to travel to the far end where it is doubled to 4.0 p.u.; this voltage is superposed on the initial $V_t = -1.0$ p.u., resulting in a voltage to ground of 3.0 p.u. as shown in *Figure 2.10*, Curve 2.

Practical cases. The simple waveshapes illustrated in *Figure 2.10* are, in practice, modified by source impedance and coupling between phases. A power station can be represented by an inductance whose reflection factor is time-dependent and changes the shape of the reflected wave. This is illustrated in *Figure 2.11*, Curves 1 and 2, for a 430 km line with a source impedance of 0.1 H, corresponding to a fault level at the 400 kV bus of 5000 MVA; the maximum overvoltages in this case were computed at 2.56 p.u. without trapped charge and 4.05 p.u. with trapped charge. Low short-circuit levels tend to increase the overvoltage. Additional lines connected to the supply bus, on the other hand, have in general a favourable effect.

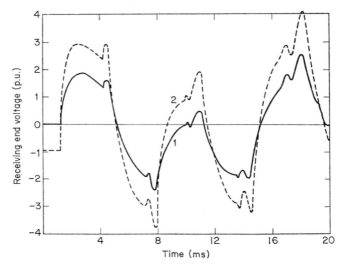

Figure 2.11. Receiving end voltage when energising a 430 km line from a 0.1 H inductive source (after Bickford and Doepel[32]):
1, without trapped charge,
2, with trapped charge

If all three poles were to close simultaneously on an ideally balanced three-phase line, this case would be identical with the single-phase case. In reality this is impossible to achieve because of pre-striking, mechanical tolerances in the switching device and the presence of unbalancing ground and ground wires. In practice, mutual effects between phases produce complicated surges which have been studied by digital computation[32, 33] and transient analyser methods[34, 35]. The overvoltage peaks are found to be higher than for the single-phase case; they depend very much on the 'pole-closing span', i.e. the time difference between the instants the first and last poles close. For open-ended lines the overvoltages decline with shorter pole-closing spans. For transformer-terminated lines the relationship is not so clear; the presence of a delta-winding, for example, affects the phase coupling. A span of 3 ms is considered good from the point of view of breaker design.

Without any special measures, apart from using modern restrike-free circuit-breakers, the maximum phase-to-ground overvoltages when energising an uncharged line are typically 2.6 p.u. but with trapped charges they can be as high as 4.5 p.u.

2.3.4 Control of energising overvoltages

The measures for the control and reduction of switching surges derive from the mechanism of their generation discussed in the preceding section. They are:

(1) One-step or multi-step energisation of transmission lines by pre-insertion of suitably dimensioned resistors.

(2) Phase-controlled closing of circuit-breakers.

(3) Drainage of trapped charges before reclosing.

(4) Use of shunt reactors.

(5) Switching surge limitation by surge diverters (see 7.2).

(1)   *Pre-insertion resistors.* The most common measure for reducing energising transients consists of applying the voltage to the line through a resistor (R) in the first stage and short-circuiting it after a brief delay[22]. The source impedance is small compared to R and can be neglected. The voltage step applied in the first stage is reduced to $Z_0/(R+Z_0)$ per unit; it returns from the far end unchanged in magnitude and is reflected at the near end with the reflection factor $(R-Z_0)/(R+Z_0)$. If, for example, $R = Z_0$, the voltage applied at the near end is 0.5 p.u., the voltage at the far end is 1.0 p.u. when the first wave arrives, and there can be no further reflections at the near end.

In the second stage, when R is short-circuited, a voltage step equal to the instantaneous resistance drop enters the line. It can be shown that short-circuiting the resistor before the initial energising wave has returned to the supply end results in the same overvoltage peak as if there had been no resistor. A usual resistor insertion time of 7 ms at 50 Hz, corresponding to 120 electrical degrees, permits two return trips on a 500 km line. Assuming a reflection factor at the sending end not too different from zero, the resistance drop will in this time approach its steady-state value. The larger the resistance, the greater will be the overvoltage generated by short-circuiting it, although the first-stage overvoltage is reduced. There must be an optimum value of resistance as can be seen from *Figure 2.12*, which shows the voltages at the receiving end of a 275 km, 60 Hz line for the two switching stages as a function of R/Z. It will be seen that in this case the minimum overvoltage of 1.25 p.u. is obtained for $R/Z = 0.53$. The overvoltage reduction and the optimum R/Z vary with line length. For a 415 km, 60 Hz line, for example, the optimum $R/Z = 0.37$ and the corresponding overvoltage is 1.38 p.u.

Conventional opening resistors have too high ohmic values to be effective for closing duty. In practice, pre-insertion resistors of suitable value reduce the maximum overvoltages to less than 2.5 and 2.0 p.u., with and without trapped charges respectively. Unloaded transformers connected to the far end change the picture moderately. Energising lines through transformers (i.e. low-voltage switching) causes over-voltages of the same order at the remote end, provided the breaker is equipped with pre-insertion resistors of a value approximately equal to the line surge impedance reflected to the low-voltage side[36]. The sending end transformer is very effective in draining trapped charges. Phase-to-phase overvoltages do not exceed 3.4 in per unit of phase-to-neutral voltage peak[37].

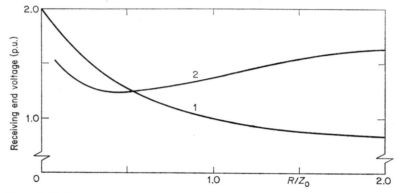

Figure 2.12. Receiving end overvoltage when energising 275 km, 60 Hz line through pre-insertion resistor:
1, resistor closing,
2, resistor shorting

Insertion times are chosen between 6 and 10 ms. The longer the insertion time the better electrically, but there is a physical limitation in the thermal capacity of the resistors. On the same grounds it is often necessary to select a higher than optimum value of the resistor and sacrifice some of the potential improvement.

For future ultra-high voltage systems, a greater surge reduction than is possible with one resistance step becomes necessary. This requirement can be met, at the cost of greater complication, by the use of multi-step pre-insertion resistors[38].

(2) *Phase-controlled switching.* The ideal solution is to eliminate the voltage step applied to the line by controlling the exact instants of closing in the three phases. This calls for a fine adjustment of the breakers, which is difficult to maintain under the large range of weather conditions to which equipment is subjected and could also be vitiated by pre-strikes. By combining an imperfect phase control with pre-insertion resistors, satisfactory results can be achieved; the maximum overvoltage is of the order of 1.5 p.u.[39].

(3) *Drainage of trapped charges.* Effective removal of trapped charges reduces the magnitudes of overvoltages on reclosing to the levels

occurring on energising. Opening resistors assist in draining trapped charges to a limited extent; their effectiveness depends on resistance value, length of line and in-circuit time. In wet conditions insulator leakage also helps[40]. Magnetic-type potential transformers discharge lines very effectively; as they oscillate with the line capacitance at low frequency, saturation sets in, causing strong damping[41]. If, however, the lines are shunt-compensated, the potential transformers cannot drain the line because of the higher natural frequency imposed by the shunt reactors which prevents saturation.

(4)   *Shunt reactors.* E.H.V. lines are frequently equipped with shunt reactor, for reasons discussed in 2.2.3. It is therefore fortunate that reactors have the incidental advantage of lowering surges caused by line energisation. This is brought about mainly by the reduction in temporary overvoltages. Reactors cannot, however, drain trapped charges which, after line disconnection, oscillate between shunt inductance and line capacitance (at a frequency which is usually somewhat below normal frequency since compensation is normally less than 100%). The decay is slow because of the high X/R ratio. *Figure 2.13a*

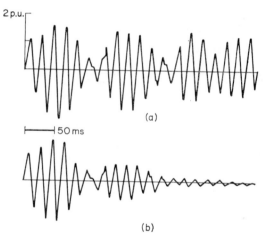

Figure 2.13. Voltage oscillations after interruption of shunt reactor compensated line:
(a) normal damping,
(b) additional resistors in series with reactors.
Note beats caused by line-mode and ground-mode oscillations.

shows such an oscillation at 40 Hz, with an overswing to 1.15 p.u. voltage. It is clear that reclosing during this oscillation at an unfavourable instant would produce a high overvoltage.

An effective way of reducing the trapped charges during the 'dead' time before reclosing is by the temporary insertion of resistors in series with the reactors; the strong damping achieved is illustrated in *Figure 2.13b*. Optimum values of pre-insertion resistors are higher when shunt reactors are installed; with correctly chosen values the maximum overvoltage on reclosing can be reduced to below 1.5 p.u.

Conclusion.   Techniques are available to keep switching overvoltages to values of the order of 2 p.u. which permit economical insulation design of systems up to 765 kV. Refinements are possible and are being developed for practical application which should lower surge levels still further, as demanded for future systems of 1100 kV and higher.

REFERENCES—CHAPTER 2

1. Bewley, L. V., *Travelling Waves on Transmission Systems*, Dover Publications, New York, N. Y., (1963).
2. Rudenberg, R., *Electric Shock Waves in Power Systems*. Harvard University Press, Cambridge, Mass., (1968).
3. Simpson, G. C., and Scrase, F. J., 'The Distribution of Electricity in Thunderclouds', *Proceedings Royal Society*, Series A, **161**, 309 (1937).
4. Schonland, B. F. J., *The Flight of Thunderbolts*, Clarendon Press, Oxford, (1964).
5. Bruce, C. E. R., and Golde, R. H., 'The Lightning Discharge', *Journal IEE*, **88, Pt. II,** 487–505 (1941).
6. Berger, K., 'Novel Observations on Lightning Discharges: Results of Research on Mt. San Salvatore', *Journal Franklin Institute*, **283**, No. 6, 478–525 (1967).
7. Westinghouse Electric Corp., *Transmission and Distribution Reference Book*, 4th edn., East Pittsburgh, Pa., (1950).
8. Prentice, S. A., 'CIGRE Lightning Flash Counter', *Electra* (CIGRE), No. 22, 149–169 (1972).
9. AIEE Committee report, 'A Method for Estimating Lightning Performance of Transmission Lines', *AIEE Transactions Pt. III*, **69**, 1187–1196 (1950).
10. General Electric Company, *EHV Transmission Line Reference Book*, Edison Electric Institute, New York, N.Y., (1968).
11. Sargent, M. A., 'The Frequency Distribution of Current Magnitudes of Lightning Strokes to Tall Structures', *IEEE Winter Meeting*, Paper No. T72, 216–225 (1972).
12. World Meteorological Organisation, *World Distribution of Thunderstormdays*, Geneva, Pt. I (1953), Pt. II (1956).
13. Muller-Hillebrand, D., Johansen, O., and Saraoja, E. K., 'Lightning Counter Measurements in Scandinavia', *Proceedings IEE*, **112**, 203–210 (1965).
14. Prentice, S. A., and Mackerras, D., 'Recording Range of a Lightning Flash Counter', *Proceedings IEE*, **116**, No. 2, 294–302 (1969).
15. Dandeno, P. L., and McClymont, K. R., 'Extra-High-Voltage System Overvoltages following Load Rejection of Hydraulic Generation', *IEEE Transactions, Power Apparatus and Systems*, **82**, 49–57 (1963).
16. DeMello, F. P., Dolbec, A. C., Swann, D. A., and Temoshok, M., 'Analog Computer Studies of System Overvoltages following Load Rejection', *IEEE Transactions, Power Apparatus and Systems*, **82**, 42–49 (1963).
17. Jancke, G., 'The Development of the Swedish 400 kV Network', *IEEE Transactions, Power Apparatus and Systems*, **83**, 197–295 (1964).

18. Dineley, J. L., and Glover, K. J., 'Voltage Effects of Capacitive Load on the Synchronous Generator', *Proceedings IEE*, **111**, 789–795 (1964).
19. Booth, R. R., 'The Stability of Alternators with a Capacitive Load', *Institution of Engineers Australia, Electrical Engineering Transactions*, **EE2**, No. 2, 88–96 (1966).
20. Willheim, R., and Waters, M., *Neutral Grounding in High-Voltage Transmission*, Elsevier, N.Y., London, (1956).
21. International Electrotechnical Commission (IEC), Technical Committee 28, Document (Central Office) 35, (1970).
22. Hedman, D. E., Johnson, I. B., Titus, C. H., and Wilson, D. D., 'Switching of Extra-high-voltage Circuits: Pt. II—Surge Reduction with Circuit-breaker Resistors', *IEEE Transactions, Power Apparatus and Systems*, **83**, 1196–1205 (1964).
23. Smith, D. C., and Bates, I. P., 'Dynamic Overvoltages in the Victorian 500 kV Transmission Network', *Institution of Engineers Australia, Electrical Engineering Transactions*, **EE4**, No. 1, 147–157 (1968).
24. Kimbark, E. W., and Legate, A. C., 'Fault Surge versus Switching Surge, A Study of Transient Overvoltages by Line-to-ground Faults', *IEEE Transactions, Power Apparatus and Systems*, **87**, 1762–1764 (1968).
25. Colclaser, R. G., Wagner, C. L., and Buettner, D. E., 'Transient Overvoltages Caused by the Initiation and Clearance of Faults on a 1100 kV System', *IEEE Transactions, Power Apparatus and Systems*, **89**, No. 8, 1744–1751 (1970).
26. White, E. L., and Reece, M. P., 'Statistical Data on Switching Surges in the British Grid', *Proceedings CIGRE*, Report 337 (1958).
27. AIEE Committee Report, 'Switching Surges, I—Phase-to-Ground Voltages', *AIEE Transactions, Power Apparatus and Systems*, **80**, 240–261 (1961).
28. IEEE Committee Report, 'Switching Surges, II—Selection of Typical Waves for Insulation Coordination', *IEEE Transactions, Power Apparatus and Systems*, **85**, 1091–1097 (1966).
29. IEEE Committee Report, 'Switching Surges, III—Field and Analyser Results for Transmission Lines, Past, Present and Future Trends', *IEEE Transactions, Power Apparatus and Systems*, **89**, 173–189 (1970).
30. Lauber, T. S., 'Resistive Energization of Transmission Lines, Pt. I—One-phase Lines', *IEEE Conference Paper*, 31CP65-174 (1965).
31. Lauber, T. S., 'Resistive Energization of Transmission Lines, Pt. II—Three phase Lines', *IEEE Conference Paper*, 31CP65-171 (1965).
32. Bickford, J. P., and Doepel, P. S., 'Calculation of Switching Transients with Particular Reference to Line Energization', *Proceedings IEE*, **114**, 465–477 (1967).
33. Dommel, H. W., 'Digital Computer Solution of Electromagnetic Transients in Single- and Multiphase Networks', *IEEE Transactions, Power Apparatus and Systems*, **88**, No. 4, 388–399 (1969).
34. Alexander, G. W., Mielke, J. E., and Trojan, H. T., 'Switching Surges on Northern States Power Company's 345 kV Circuits', *IEEE Transactions, Power Apparatus and Systems*, **88**, No. 6, 919–931 (1969).
35. Hauspurg, A., Vassell, G. S., Stillman, G. I., Charkow, J. H., and Haahr, J. C., 'Overvoltages on the AEP 765 kV System', *IEEE Transactions, Power Apparatus and Systems*, **88**, 1329–1342 (1969)
36. Essel, C. J., Low, S. S., Boehr, J. W., and Reed, N. E., 'Field Tests of Low-side Switching of EHV Transmission Line', *IEEE Transactions, Power Apparatus and Systems*, **90**, 1492–1503 (1971)
37. Wilson, D. D., 'Phase-Phase and Phase-Neutral Switching Surges on 500 kV Open-ended Lines', *IEEE Transactions, Power Apparatus and Systems*, **88**, No. 5, 660–665 (1969).
38. Colclaser, R. G., Wagner, C. L., and Donohue, E. P., 'Multi-step Resistor Control of Switching Surges', *IEEE Transactions, Power Apparatus and Systems*, **88**, 1022–1028 (1969).

39. Thorén, H. B., 'Reduction of Switching Overvoltages in EHV and UHV Systems', *IEEE Transactions, Power Apparatus and Systems*, **90,** 1321–1324 (1971).
40. Beehler, J. E., 'Weather, Corona, and the Decay of Trapped Energy on Transmission Lines', *IEEE Transactions, Power Apparatus and Systems*, **83,** 512–520 (1964).
41. Marks, L. W., 'Line Discharge by Potential Transformers', *IEEE Transactions, Power Apparatus and Systems*, **88,** 293–299 (1969).
42. Clerici, A., Ruckstuhl, G., and Vian, A., 'Influence of Shunt Reactors on Switching Surges', *IEEE Transactions, Power Apparatus and Systems*, **89,** No. 8, 1727–1736 (1970).

3

Disruptive discharge and withstand voltages

3.1 INTRODUCTION

In insulation co-ordination the concept of a 'withstand voltage' is fundamental. If it could be defined as a precise voltage which, if exceeded, would definitely lead to insulation breakdown and when not exceeded would be 100% safe, and if the maximum overvoltage were equally well known, the matching of withstand voltage to maximum overvoltage would be easy: one would simply choose an adequate safety factor as the ratio of these two voltage values. This 'conventional' method is indeed used, but as there is neither a precise value of withstand voltage nor, very often, of maximum overvoltage, it is necessary to adopt large 'factors of ignorance' which become very costly at extra-high voltages. The withstand voltage level is properly defined on a statistical basis as the voltage at which the probability of a disruptive discharge has a certain, suitably low, value[1]; it is thus tied to the mean disruptive discharge voltage. Combining this conception with the statistical distribution of overvoltages, one arrives logically at a probabilistic treatment of breakdown. The result is a statement concerning the probability of failure or the probable number of flashovers in a given period. Only elementary statistical theory is required and will be included in this chapter.

The approach differs for self-restoring and non-self-restoring insulation. The former is generally external insulation which recovers the integrity of its insulating properties after disruptive discharges. It can therefore be subjected to repeated flashovers for the purpose of gaining statistical information. This is not so with non-self-restoring insulation which is usually internal insulation comprising solid, liquid or gaseous elements of equipment insulation protected from atmospheric influences.

3.2   SELF-RESTORING INSULATION

3.2.1   Statistical impulse withstand voltages

For self-restoring insulation a flashover characteristic can be determined by a large number of impulse voltage applications with crest values varying over a suitable range but with identical shape and polarity. One method would be to apply, say, 20 voltage impulses at each of a number of voltage levels, the voltage increasing in small steps[2]. The number of discharges at each level, divided by the number of voltage applications, is the approximate probability of discharge (p) for the particular impulse shape, magnitude, and polarity. To obtain accurate values, the number of applications would have to be very large.

If flashover probability (p) is plotted against voltage and an average curve drawn, a 'frequency distribution' curve results. It conforms quite closely to a 'normal' or 'Gaussian' distribution, shown as Curve 1 on *Figure 3.1a*. When plotted on 'normal probability paper' (paper with one scale suitably distorted), this curve turns into the straight line shown on *Figure 3.1b*. This fact can be used to test whether a particular distribution conforms to a normal one.

The point for $p = 0.50$ fixes the '50% disruptive discharge voltage' or 'critical flashover (CFO) voltage'. Alternative methods for determining $\overline{V}$, which require fewer test points, are described in [3] (Clauses 10.3 and 10.4).

For the normal distribution[4], the probability of discharge at any voltage (V) can be represented by equation 3.1 in which the variable (V) has been transformed into the normalised variable (z):

$$p = \int_{-\infty}^{z} \frac{1}{\sqrt{2\pi}} \exp\left(-\frac{1}{2}z^2\right) dz \tag{3.1}$$

where

$$z = (V - \overline{V})/\sigma \tag{3.2}$$

$$\overline{V} = \Sigma V/n \tag{3.3}$$

n = number of tests

$$\sigma = [\Sigma(V - \overline{V})^2/n]^{1/2} \tag{3.4}$$

The 'standard deviation' σ is a measure of the scatter or dispersion of the observations (V) about the mean or CFO voltage ($\overline{V}$). $\sigma/\overline{V}$ is called the coefficient of variation (or per unit standard variation).

29

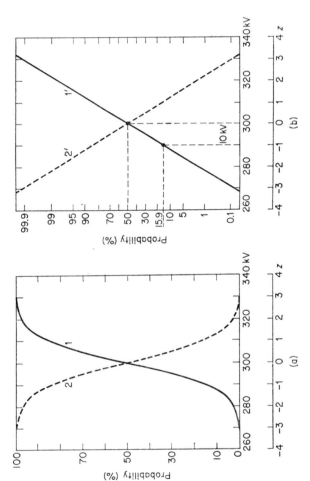

Figure 3.1. Cumulative normal distribution functions:
1, 1', probability of disruptive discharge,
2, 2', probability of withstand.
(a) linear scales,
(b) plotted on normal probability paper.

Note that abscissa can be in kV or in the normalised variable $z = (V - \overline{V})/\sigma$. Illustrated: $\overline{V} = 300 \text{ kV}$, $\sigma = 10 \text{ kV}$.

The advantage of the normal distribution is that, once the CFO voltage and the standard variation are known, the probability of a disruptive discharge can be estimated for any voltage. *Table 3.1* gives some useful numerical values for equation 3.1.

Table 3.1 NORMAL DISTRIBUTION

z	-3.0	-2.0	-1.28	-1.0	0	1.0	1.28	2.0	3.0
p	0.0013	0.023	0.10	0.159	0.500	0.841	0.90	0.977	0.9987

From *Table 3.1* it is seen that, if the voltage is by twice the standard deviation below the critical voltage ($z = -2$), the probability of a disruptive discharge $p = 0.023$ (or 2.3%) and the probability of withstand $1-p = 0.977$. In this manner, the withstand voltage can be given a perfectly clear meaning.

In recent proposals[5], the International Electrotechnical Commission defines a 'reference withstand probability' and suggests for it a numerical value of 0.90 (hence $z = -1.28$). It is difficult to prove probabilities for large departures from the mean, because of the large number of test points that would be needed. It is therefore recommended that the withstand voltage should be verified by testing for the CFO voltage which must be at least equal to $1/(1-1.28\sigma/\overline{V})$ times the stipulated withstand voltage ($\sigma/\overline{V}$ = coefficient of variation).

In some cases a larger margin between CFO and statistical withstand voltages is adopted, e.g. $z = -2$ to $z = -3$.

Approximate figures for the coefficient of variation of self-restoring insulation, which can be used when more detailed information is unavailable, are:

3% for lightning impulse voltage, dry or wet, and power frequency voltage dry,

6% for switching impulse voltage, dry or wet, and power frequency voltage, wet.

Liquid and solid insulation have higher coefficients of variation, e.g. 10% for transformer oil, 8% for resin bonded paper.

Withstand and discharge are mutually exclusive events; together they add up to the totality of events in the test. It follows that the sum of the two probabilities must be unity and the probability of withstand is $q = 1-p$. The curves 2 and 2' in *Figure 3.1* represent the distribution of the withstand voltages.

3.2.2 Statistical a.c. or d.c. withstand voltage

For power frequency and d.c. voltages the test procedure described in 3.2.1 has to be modified. The voltage is slowly raised until a disruptive discharge occurs, and this is repeated a number of times. The CFO voltage is the mean of the measured values (equation 3.3) and the standard deviation is obtained from equation 3.4 or, if n is small, from

$$\sigma = [\Sigma(V - \bar{V})^2/(n-1)]^{1/2} \tag{3.5}$$

3.2.3 Confidence limits

The probability distribution derived from a limited test series departs more or less from the 'true' distribution, depending on the number (n) of tests at each voltage level. It is possible to estimate by statistical methods the limits between which the true values of mean and standard deviation will lie with a given probability. In the field of discharge voltages, this probability is usually chosen as 0.95 and one speaks of the '95% confidence limits'. Their approximate ranges are[3]

$$\text{for CFO voltage} \qquad \pm t_p \sigma / \sqrt{n}$$
$$\text{for standard deviation} \quad \pm [t_p/(n-1)]^{1/2}\sigma \tag{3.6}$$

Values of t_p can be taken from textbooks on statistics. For a 95% confidence level, $t_p \simeq 2$ and the confidence limits for $n = 20$ are approximately

for CFO voltage $\pm 0.45\sigma$
for standard deviation $\pm 0.33\sigma$

For $\sigma = 6\%$, as for switching surges, the CFO can be determined with a good accuracy of $\pm 3\%$, but the standard deviation has a very unsatisfactory range of $\pm 33\%$. Because of this difficulty one often resorts to the use of values obtained by combining the results of similar tests in many laboratories.

3.2.4 The probability density function

In Chapter 2, the frequency distribution function was used to present information on switching overvoltages. These conform closely to a normal distribution, as can be seen from *Figure 2.9*. The statistical concepts developed for withstand voltage can be readily applied to the definition of probable maximum overvoltages.

For the discussion of outage risks in Chapter 6 it will be convenient to use the probability density function (p_0) which is derived from equation 3.1 by differentiation:

$$p_0 = (1/\sqrt{2\pi}) \exp(-\tfrac{1}{2}z^2) \tag{3.7}$$

The physical interpretation of this function (*Figure 3.2*) is as follows: the probability that a quantity (V) will have values between V_1 and $V_1 + dV$ is given by $p_0 dz$ for the corresponding abscissa z_1 and interval dz. By integrating from $-\infty$ to z_1 one obtains in the crosshatched area under the p_0 curve the probability that all quantities below the value V_1 (z_1) occur. By plotting the integral values, one arrives again at the frequency distribution or cumulative probability function. The integral from $-\infty$ to $+\infty$ is unity, i.e. certainty that all values are included.

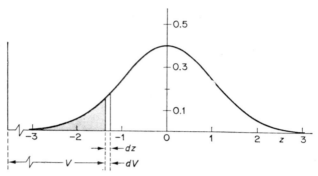

Figure 3.2. Probability density function

3.2.5   Impulse withstand tests

The method outlined in 3.2.1 is unsuitable for test objects combining self-restoring with non-self-restoring insulation. American Standards[6] prescribe for such cases the application of three impulses at the withstand level. If all three are withstood, the test has been passed; if two or three flashovers occur the test has failed, but if only one flashover occurs (on the self-restoring insulation), a second series of three impulses is applied with no flashover allowed. It is interesting to calculate the overall probability q_w of passing this test as a function of the probability q of withstand for each separate application. *Table 3.2* lists the combined probabilities* of the four possible satisfactory events. (F stands for flashover and W for withstand. Probability of flashover: $p = 1 - q$.)

*The following rules for combining probabilities are recalled:
(1) The probability that all of n independent events will occur is the product of the n individual probabilities. (Example: probability of heads on both of two tosses of a coin is $\frac{1}{2} \cdot \frac{1}{2} = \frac{1}{4}$.)
(2) The probability of one or other of n mutually exclusive events occurring is the sum of the n individual probabilities. (Example: probability of a 2 *or* a 6 on one toss of a die is $1/6 + 1/6 = 1/3$.)

Table 3.2 CALCULATION OF PROBABILITY OF PASSING STANDARD TEST

Event	Probability
W—W—W	q^3
F—W—W...W—W—W	$(1-q)q^5$
W—F—W...W—W—W	$q(1-q)q^4$
W—W—F...W—W—W	$q(1-q)q^3$

The overall probability of passing the whole test is the sum of the probabilities of all satisfactory events:

$$q_w = q^3 + 3(1-q)q^5 \qquad (3.8)$$

The graph q_w v. q in *Figure 3.3* shows that $q = 90\%$ gives an overall $q_w = 92\%$. This is not a stringent test for important equipment. IEC in proposed amendments[7] to Publication 60 requires the application of 15 impulses with two disruptive discharges on insulation allowed. The second graph in *Figure 3.3* illustrates the much greater severity of this test.

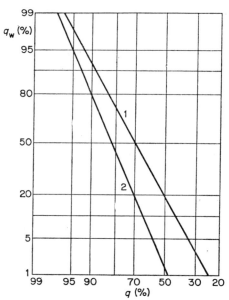

Figure 3.3. Probability q_w of passing test if the test object has a probability q of withstanding a single test:

1, ANSI standard,
2, IEC standard

3.2.6   Environmental effects

The strength of external insulation depends on air density, humidity, precipitation and contamination. It is referred to a standard reference atmosphere of temperature 20°C, pressure 1013 mbar (760 mm Hg at 0°C) and absolute humidity of 11 g/m³.

The disruptive discharge voltage increases with air density and humidity; the former reduces the mean free path of charge carriers and the latter reduces their mobility as they are captured by water molecules. The disruptive discharge voltage is related to that at standard atmosphere by the factor k_d^m/k_h^n (IEC[8]).

Relative air density

$$k_d = 0.289b/(273 + \theta) \tag{3.9}$$

where b is barometric pressure in millibar and θ the temperature in °C. The humidity factor k_h is expressed in graphs and the exponents m and n in tabular form in[8]. These parameters depend on the type and polarity of the voltage and the length of the gap. Their numerical values are stated in different ways in national standards; they are continually being improved by new test information, e.g. see[9,10].

Rain reduces the disruptive discharge voltage very considerably for power frequency and switching impulses but only little for lightning impulses. The power frequency and switching impulse values for outdoor insulation are accordingly measured in wet tests, while for lightning impulses only dry tests are made. As an indication of the magnitude of the rain effect, the power frequency CFO of a vertical insulator string in per unit of the dry value is 0.8 at a precipitation rate of 1.5 mm/min and 0.75 for a rate of 3 mm/min.

Contamination is caused by a large variety of agents (coal and cement dust, fly ash, salt spray, fertiliser, etc.) which, when moistened by fog or light misty rain, can reduce the power frequency flashover voltage of porcelain insulation to half or even to a quarter, depending on the type and deposition density of the contaminant, and the frequency of washing rain. Within the frame of this book it is not possible to deal with this topic in any detail, and the reader is referred to the literature[11-13]. An approximate guide to insulation design for contaminated conditions is the leakage distance per kV r.m.s. of the line–line operating voltage; it varies from 15 to 30 mm for light and heavy pollution, respectively. For standard cap and pin suspension insulators, the leakage distance is twice the spacing; for fog-type, a factor of three is achievable.

The switching impulse flashover voltage is approximately twice the power frequency peak flashover voltage for the same contamination condition, assuming that the insulator is not simultaneously energised at power frequency. Under combined stress, design for power frequency is normally adequate, provided the switching surge does not exceed about 1.9 times the normal power frequency peak voltage. Lightning impulses are less critical and air gaps are entirely unaffected by contamination.

3.2.7 Test values

A large amount of test information has been accumulated over the years. As operating voltages increase, experimental results are added for ever longer gaps and insulator strings[14-20]. Voltage limits imposed by the dimensions of the laboratory or the capacity of the test equipment are removed by the construction of new laboratories and by open air testing. Wet tests suffer from the difficulty that uniform artificial rain is hard to achieve, and random errors are therefore large.

Different insulation arrangements have different characteristics; this adds enormously to the test load. The most important arrangements required in practice are:

rod–rod gaps, used as protective gaps (also applicable to phase-to-phase clearances);

rod–plane gaps, representing a live conductor above ground;

parallel gaps, air clearance between phases or phase conductors and ground wires;

insulator strings for transmission lines, vertical, horizontal, vee-strings, disc and long-rod, fog type, etc.;

support insulators of various shapes.

For e.h.v., the proximity of grounded tower metal affects the flashover of insulator strings and has been studied in tests of model rigs with the objective of obtaining the optimum relationship of string length and tower clearance.

Confronted with any particular design problem, the engineer will first search the literature for test results which may fit his case; if unsuccessful and the case is sufficiently important, special tests may have to be conducted.

3.2.8 Lightning impulse test characteristics

The international standard waveshape has a 1.2 μs front and a time to half value of 50 μs (*Figure 3.4*). Waveshapes of actual surges vary considerably; it is therefore fortunate that moderate deviations from the standard full wave have little effect on the flashover voltages provided the shape remains similar. The relevant tolerances on the IEC standard have therefore been made ample (±20% on front time, ±30% on time to half value), tolerance of peak value is ±3%. The US standard wave 1.5/40 μs falls within these tolerances.

Positive polarity impulses give somewhat lower flashover values than negative ones. The effect of rain is slight, tests are therefore conducted dry. The flashover values are substantially proportional to gap length.

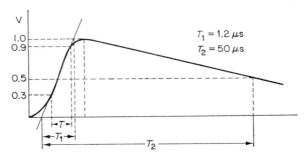

Figure 3.4. Standard impulse test wave 1.2/50 μs

As the applied impulse crest is increased for a particular test arrangement, the time to flashover falls. The instant of flashover moves from the tail of the wave to the crest and ultimately to the front of the wave (*Figure 3.5*). For times of less than about 10 to 16 μs, depending

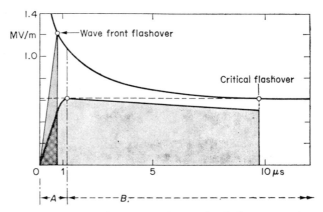

Figure 3.5. Volt-time characteristic for impulse flashovers; time ranges:
A, wave front flashover,
B, wave tail flashover

on crest magnitude, the flashover voltages increase; the increase becomes substantial for times below 5 μs, as shown in *Figure 3.5*. This shape of the volt/time characteristic is of great significance in the case of insulation subjected to short duration surges.

Figures 3.6 and *3.7* reproduce typical spark-over characteristics for rod gaps and standard suspension insulators. Positive polarity values

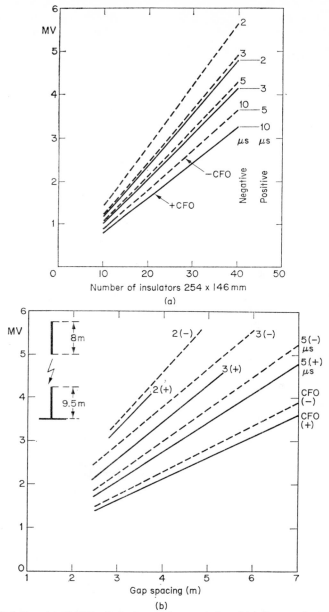

Figure]3.6. Impulse (1.2/50 μs) flashover characteristics of (a) long insulator strings (254×146 mm units) and (b) long rod gaps, at various times to flashover, corrected to standard atmospheric conditions (after Udo[17])

are given in *Figure 3.7* because they are the more critical. Information on insulation strength of other types and dimensions of insulators and gaps and on other polarities can be found in the literature; e.g.[21] gives in Section 8.8 curves synthesised from the work of several modern investigators.

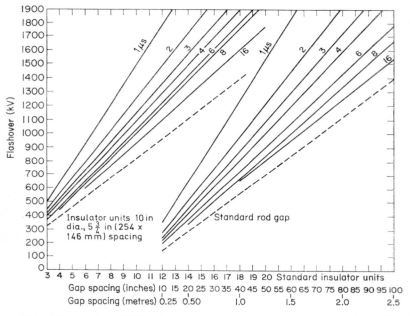

Figure 3.7. Impulse (1.2/50 μs) flashover characteristics of insulator strings (254 × 146 mm units) and rod gaps, at various times to flashover, corrected to standard atmospheric conditions (after Westinghouse Corp.[22]).
Broken lines: power frequency dry flashover peak values.

3.2.9   Switching impulse test characteristics

The large variety of switching surge waveshapes and the correspondingly large range of flashover values make it difficult to choose a standard shape of switching impulse. The tendency has been to employ an impulse waveshape that is thought to lead to the lowest flashover and withstand voltage values. In many tests the most onerous impulses are unidirectional, of positive polarity, and have a time to crest in the range from 100 to 500 μs. The time to half value has less effect because flashover almost always takes place before or at the crest.

For testing purposes current standards prescribe double-exponential shapes similar to the standard lightning impulse wave but with much longer fronts and tails. One merit of this choice is that the waves are reproducible in the laboratory. Not enough is known about the withstand characteristics of oscillatory and composite shapes. More attention is now being given to these but it will take a large effort to acquire the needed information for the whole field of switching surges.

The standard switching impulse recommended by IEC[5] has a 250 μs front time and a time to half value of 2500 μs, with alternative shapes

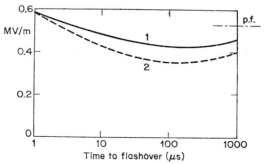

Figure 3.8. Typical relationships for critical flashover voltage per metre as a function of time to flashover (3 m gap):

1, rod–rod gap; 2, conductor–plane gap;
p.f., power frequency CFO

of 100/2500 μs and 500/2500 μs. The tolerance on peak value is ±3%, on front time ±20%, and on time to half value ±60%. *Figure 3.8* illustrates the effect of time to crest on flashover values. Lightning impulses give the highest values whilst switching impulses fall to a minimum which is below power frequency flashover voltage. In *Figure 3.9* the comparative effect of increasing electrode distance on

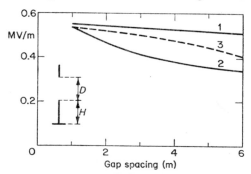

Figure 3.9. Typical relationships for flashover voltage per metre as a function of gap spacing:

1, 1.2/50 μs impulse,
2, 200/2000 μs impulse (rod–rod, $H/D = 1.0$, positive dry),
3, power frequency

power frequency and switching surges is shown by typical relationships. The drastic reduction of the average switching impulse strength per metre with increasing distance is apparent. In the ultra-high voltage range, this 'saturation' effect leads to very large and costly design clearances. It therefore becomes important to eliminate unnecessary margins.

There are four major influences on switching surge strength: geometry, the weather, statistical variations, and waveshape.

(1)   *Geometrical factors.*   The shape of both electrodes affects the formation and propagation of streamers in large gaps; this explains the influence on switching surge flashover of corona shields and bundle conductor configuration on the one hand, and the shape and physical size of the towers on the other. It has been found, for example, that small lightweight towers have a greater flashover strength than large heavy ones. It is necessary for any new tower and insulator configuration, for which test values are not available, to co-ordinate string

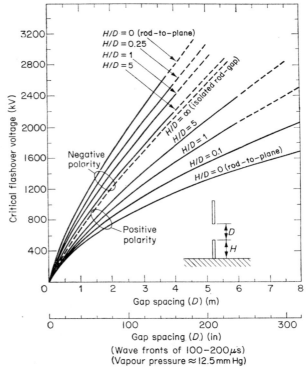

Figure 3.10. Switching surge flashover strength of rod–rod and rod–plane gaps (courtesy Edison Electric Institute[21])

length and tower clearances by experiment in order to obtain an economical solution.

Figure 3.10 illustrates the very large proximity effect of the ground plane on the switching surge flashover voltages of both polarities. The extreme curves for $H/D = 0$ represent the positive and negative rod-to-plane flashover strength; intermediate curves show the gradual transition between these two cases. The dashed curve for $H/D = \infty$ is an interpolation corresponding to the case of an isolated rod–rod gap; if it could be realised physically it should show independence from polarity effects.

A useful qualitative explanation by Anderson and Tangen[21] is as follows:

'Breakdown in non-uniform field electrode configurations is dominated by a positive-polarity streamer process.... When the upper electrode is positive, the positive polarity streamers are formed at that electrode ... the oppositely charged ground plane ... enhances the gradients at the streamer tips.... The closer the ground plane is moved to the gap D, the more it enhances the positive breakdown streamers from the upper electrode, making the configuration weaker until the final position of the ground plane at the tip of the lower electrode (rod-to-plane) is weakest of all.

Conversely, for the upper rod negative ... the positive-polarity streamers have to form off the lower rod, which is now of the same polarity as the adjacent ground plane.... As the ground plane is moved upward (H approaching zero), the lower rod ... completely disappears into the ground plane and the positive streamers—if they form at all—are forced to form in the weak, almost uniform field at the ground plane ... only the less active negative streamers can accomplish the bridging of the gap. Therefore the negative-polarity strength of gap increases as the ground plane comes closer to it and shields more of the positive-polarity rod'.

A practical tower configuration will have a flashover voltage corresponding to the curve in *Figure 3.10* approximating most closely its effective H/D ratio (see Chapter 6).

'Phase-to-phase switching surge strength' is a complicated phenomenon because, as TNA studies have shown, there can be a time displacement between the surges on parallel conductors, and the ratio of the positive and negative voltages on the two electrodes may vary. The time displacement causes the maximum voltage between phases to be less than the sum of the phase–ground voltages. In experimental tests it has been shown, for example, that if the positive surge leads the negative surge, the CFO is increased, and in the reverse case the CFO is decreased. This matter requires considerable exploration before reliable design criteria can be evolved. In the meantime, the guide lines given in[21] Chapter 6 can be accepted for practical purposes.

(2)   *Meteorological factors* were discussed in 3.2.6.

(3)   *Statistical fluctuations in switching surge strength.* Even after correcting for all known factors it is found that a considerable degree of scatter exists, which is Gaussian in character, with a per unit standard deviation from 2 to 10%. In the average, a figure of 6% can be adopted.

(4)   *Waveshape.*   In order to limit the amount of testing, most experimental work in the past has used a unipolar waveshape resulting in minimum flashover values. For the future voltage levels of 1000 kV and above, this assumption may lead to unknown and unnecessary margins. Attention is currently being paid to composite waveshapes.

3.2.10   Power frequency test characteristics

Dry power frequency flashover voltages for gaps and vertical insulator strings are shown in *Figures 3.6* and *3.7*. The dry values chosen are usually very ample, in order to allow for the drastic drop for rain and contamination as discussed in 3.2.6.

3.2.11   Gaps or strings in parallel

If the probability of withstand of a single insulator string or gap for a voltage V is q_1, then the probability of withstand of n equal strings or gaps is

$$q_n = q_1^n \tag{3.10}$$

This theoretical result has been confirmed by experiment on transmission lines where numerous insulator strings in parallel are subjected to approximately the same switching surge stresses. The withstand strength of the whole line is reduced and can be estimated by the above formula. Allowance should be made for attenuation along the line.

3.3 NON-SELF-RESTORING INSULATION

3.3.1   Conventional withstand voltage

It is not practicable to test non-self-restoring insulation in valuable equipment like transformers on a statistical basis; consequently it is impossible to establish a statistical withstand voltage to which a failure probability could be attributed. All that can be done is to specify, in accordance with accepted standards, a small number of voltage applications of lightning and/or switching impulses, as the case may be. A successful test gives no quantitative indication of the withstand probability. Considerable reliance has to be placed on the fact that similar equipment by the same or other reputable manufacturers, having passed identical tests in the past, performed satisfactorily in service. Manufacturers perform tests on components of the insulation structure (to which statistical methods are often applied), and take other measures to avoid failure at acceptance test or under service conditions. If design and manufacturing methods have resulted in a satisfactory product in the past, this can be counted as additional evidence but it cannot be anything but a qualitative assessment.

3.3.2   Standard insulation levels

Current standard specifications, such as IEC Publication 71-1967[1], prescribe power frequency and 1.2/50 μs impulse withstand voltages and general testing procedures to which detailed equipment standards should adhere. These standards apply also to self-restoring insulation if it is part of equipment (bushing, circuit-breakers, disconnects, etc.). Equipment standards provide for additional tests, e.g. the 'chopped wave' test for transformers which consists of the application of a standard impulse wave, with a peak of 1.15 times the standard impulse peak, which, after 2–5 μs, is suddenly reduced to practically zero by sphere gap flashover. The negative voltage step causes severe stresses inside the winding.

Table 3.3 gives specification values of test voltages for selected operating voltages in the upper range. 'Full' insulation refers to original standardised values. In parallel with the gradual improvement of surge diverter protection (Chapter 7) the 'reduced' levels were introduced.

The conventional power frequency test for equipment has been under attack for its lack of realism; equipment is never stressed by a.c. voltages of the order of two to three times the phase voltage. In its defence it has been said that it is an easy test to perform and that equipment built to withstand it has proved satisfactory in operation. For the extra high voltages, however, unnecessary margins cannot be tolerated and this is recognised by IEC in its latest recommendations[5], in which the power frequency test is replaced by a switching surge test (SIL) for operating voltages from 300 kV upwards, as indicated in *Table 3.4*. This system will probably be extended to 245 kV and later to still lower voltages. It is likely that ultimately a power frequency test at reduced level (of the order of 1.5 p.u.) will be retained. It will be noted that the new BILs are generally lower than those of *Table 3.3*, thus reflecting the progress made in surge diverter design.

Table 3.3 EXTRACT FROM IEC PUBLICATION 71-1967[1]

Highest voltage for equipment	Impulse withstand test voltage with standard full wave positive and negative polarity		Power frequency withstand test voltage with respect to earth under standard conditions	
kV r.m.s.	Full insulation kV peak	Reduced insulation kV peak	Full insulation kV r.m.s.	Reduced insulation kV r.m.s.
145	650		275	
		550		230
		450		185
245	1050		460	
		900		395
		825		360
		750		325
362		1300		570
		1175		510
		1050		461
420		1675		740
		1550		680
		1425		630
		1300		570
525		1800		790
		1675		740
		1550		680
		1425		630

A feature of both tables is that for each 'highest voltage for equipment' a range of insulation levels is provided. In special cases it is also possible to choose insulation levels outside the given range. This may be necessary because the optimum technical and economical choice of the insulation level depends on factors which vary from system to system, even with the point of installation in the same system, and the method chosen to control switching and lightning overvoltages (see Chapter 7).

Table 3.4 STANDARD INSULATION LEVELS FOR $U_m \geqslant 300$ kV[5]

1	2	3	4	5	6
Highest voltage for equipment U_m (r.m.s.)	Base for p.u. values $U_m \dfrac{\sqrt{2}}{\sqrt{3}}$	Rated switching impulse withstand voltage		Ratio between lightning and switching impulse rated withstand voltages	Rated lightning impulse withstand voltage
kV	kV (peak)	p.u.	kV (peak)		kV (peak)
300	245	3.06	750	1.13	850
				1.27	950
		3.45	850	1.12	
362	296	2.86		1.24	1050
		3.20	950	1.12	
420	343	2.76		1.24	1175
				1.12	
		3.06	1050	1.24	1300
525	429	2.45		1.12	
				1.36	
		2.74	1175	1.21	1425
				1.10	
				1.32	
		2.08	1300	1.19	1550
				1.09	
				1.38	
765	625	2.28	1425	1.28	1800
				1.16	
				1.26	1950
		2.48	1550	1.47	2100
				1.55	2400

REFERENCES—CHAPTER 3

1. International Electrotechnical Commission (IEC), *Insulation Co-ordination*, Publication 71, 4th edn. (1967). See also proposed revisions in documents of Technical Committee No. 28 (Central Office) 35, 35A (1970), 37 (1971).
2. International Electrotechnical Commission, Technical Committee 42, *High-Voltage Testing Techniques*, Publication 60 (1962), and proposed revisions in documents (Central Office) 14 (1970), 15 (1970), 19 (1972), 20 (1972).
3. International Electrotechnical Commission, Technical Committee 42, Document 42, (Central Office) 15 (1970).
4. Spiegel, M. R., *Theory and Practice of Statistics*, Schaum's Outline Series, McGraw-Hill, New York, (1961).

5. International Electrotechnical Commission, Technical Committee No. 28, Document 28 (Central Office) 35 (1970).
6. American Standard ANSI, *Measurement of Voltage in Dielectric Tests*, C68.1 (1968).
7. International Electrotechnical Commission, Technical Committee 42, Document 42 (Central Office) 20 (1972).
8. International Electrotechnical Commission, Technical Committee 42, Document 42 (Central Office) 19 (1972).
9. Standring, W. G., Browning, D. H., Hughes, R. C., and Roberts, W. J., 'Effect of Humidity on Flashover of Air-Gaps and Insulators under Alternating (50 Hz) and Impulse (1/50 μs) Voltages', *Proceedings IEE*, **110**, No. 6, 1077–1081 (1963).
10. Harada, T., Aihara, Y., and Aoshima, Y., 'Influence of Humidity on Lightning and Switching Impulse Flashover Voltages', *IEEE Transactions, Power Apparatus and Systems*, **90**, No. 4, 1433–1442 (1971).
11. Forrest, J. S., Lambeth, P. J., Kucera, J., Colombo, A., Hurley, J. J., Limbourne, G. J., and Nakajima, Y., 'International Studies of Insulator Pollution Problems', *Proceedings CIGRE*, Report 33–12 (1970).
12. Lambeth, P. J., 'Effect of Pollution on High-Voltage Insulators', *Proceedings IEE, (IEE Reviews)* 118, No. 9R, 1107–1130 (1971).
13. Lushnicoff, N. L., and Parnell, T. M., 'The Effects of Pollution and Surface-Discharges on the Impulse Strength of Line Insulation', *IEEE Transactions, Power Apparatus and Systems*, **90**, No. 4, 1619–1627 (1971).
14. Rohlfs, A. F., Fiegel, H. E., and Anderson, J. G., 'The Flashover Strength of E. H. V. Line and Station Insulation', *AIEE Transactions Pt. III (Power Apparatus and Systems)*, **80**, 463–470 (1961).
15. Standring, M. A., Browning, D. N., Hughes, R. C., and Roberts, W. J., 'Impulse Flashover of Air-Gaps and Insulators in the Voltage Range 1–2.5 MV', *Proceedings IEE*, **110**, No. 6, 1082–1088 (1963).
16. Alexander, D. E., and Boehne, E. W., 'Switching Surge Insulation Level of Porcelain Insulator Strings', *IEEE Transactions, Power Apparatus and Systems*, **83**, 1145–1157 (1964).
17. Udo, T., 'Sparkover Characteristics of Large Gap Spaces and Long Insulator Strings', *IEEE Transactions, Power Apparatus and Systems*, **83**, 471–483 (1964).
18. Udo, T., 'Switching Surge and Impulse Sparkover Characteristics of Large Gap Spacings and Long Insulator Strings', *IEEE Transactions, Power Apparatus and Systems*, **84**, 304–309 (1965).
19. Watanabe, T., 'Switching Surge Flashover Characteristics of Extremely Long Airgaps and Long Insulator Strings', *IEEE Transactions, Power Apparatus and Systems*, **86**, 933–936 (1967).
20. Udo, T., Tada, T., and Watanabe, Y., 'Switching Surge Sparkover Characteristics of Extremely Long Airgaps and Long Insulator Strings under Non-standard Conditions', *IEEE Transactions, Power Apparatus and Systems*, **87**, 361–367 (1968).
21. General Electric Company, *EHV Transmission Line Reference Book*, Edison Electric Institute, New York, (1969).
22. Westinghouse Electric Corp., *Transmission and Distribution Reference Book*, 4th edn., East Pittsburgh, Pa., (1950).

4

Lightning overvoltages on transmission lines

4.1 INTRODUCTION

Lightning surges on transmission lines can arise by several mechanisms. The least harmful are the voltages induced by strokes to ground in the vicinity of a line. Lightning strokes to the phase conductors produce the highest overvoltages for a given stroke current. An approximate value of the conductor potential at the point of strike is easily computed. The stroke current magnitude (I) is little affected by the value of the terminating impedance (Z) which, in this case, is half the line surge impedance (Z_0) since the injected current will flow in both directions. Hence

$$V = \tfrac{1}{2} I Z_0 \tag{4.1}$$

A stroke current as low as 10 kA, which according to *Figure 2.2* has at least a 65% probability of being exceeded, will cause a potential of 2500 kV, for a typical $Z_0 = 500\,\Omega$. Only lines insulated for ultra-high voltages could withstand this stress. It is clear that most strokes to a phase conductor will cause a flashover.

Circuit interruptions following flashovers can be avoided or at least mitigated by arc suppression coils[7] or 'protector tubes'[8], devices which act in different ways to achieve the same objective, extinction of power follow arcs. On wood pole lines, a similar result can be achieved by designs which utilise the arc quenching property of wood. Finally, automatic reclosing can be used to reduce the number of sustained interruptions. Wood pole lines and reclosing will be discussed in Chapter 5.

On important lines of high lightning incidence, it is accepted practice to seek to prevent direct strokes to phase conductors by arranging one or more shielding wires (overhead ground wires) above the phase conductors to intercept lightning strokes and conduct them to ground. The potential created by current flowing through grounded parts of the line is much lower than that due to direct strokes; however, given a large enough current it can be sufficient to cause 'backflashovers'. For simple calculations of this case, discussed in 4.4 and 4.5, the equivalent circuit of *Figure A.1c* of Appendix A1 is adapted as shown in *Figure 4.1*. The stroke current (I) is derived from a constant-current

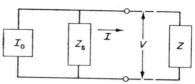

Figure 4.1. Equivalent Thevenin-circuit for lightning stroke:
I_0 = stroke current to zero-earth (constant current source);
I = stroke current to terminating impedance (Z);
Z_s = surge impedance of stroke channel;
V = voltage at stricken object

source (I_0), paralleled by the surge impedance (Z_s) of the lightning channel. The current (I) is injected into the terminating impedance (Z), representing the power system at the point of strike. It produces at the component struck a potential

$$V = IZ = I_0 Z/(1+Z/Z_s) \qquad (4.2$$

The stroke channel impedance is not well known. Calculations by various researchers give values between 1000 and 3000 Ω. As Z is small relative to Z_s, the stroke current (I) will differ only little from the stroke current to zero-resistance earth (I_0) for which statistics, such as *Figure 2.2*, are available.

These elementary calculations will give a good insight into the surge response of a transmission line and yield results of the right order of magnitude. More accurate techniques are discussed in 4.6.

A question of fundamental importance now has to be answered: how many lightning strokes will hit a transmission line or substation in a year? A knowledge of groundflash density (N_g) and of the 'area of attraction' would produce this information. It is reasonable to assume that the area of attraction will depend on the height of the most exposed conductors. Various empirical estimates of the attractive width of transmission lines have been made[1]. A simple formula gives attractive width as $4h+b$, where h is the height of the most exposed conductors on each side and b their separation.

The lack of precise data on groundflash density (equation 2.1 still allows a 1 : 2 variation) led to studies relating the number of strokes to line per annum (N_l) directly to thunderstorm level on the basis of field experience. One such formula derived from American field records[2] for lines having average-height towers of about 30 m and used in conjunction with the AIEE method of Chapter 5 is

$$N_l = 62(TD)/30 \text{ strokes}/100 \text{ km years} \qquad (4.3)$$

Another formula, based on Russian field experience[3] and applicable to towers of 25–30 m height, takes into account the average height $(\bar{h})$ of the overhead ground wire:

$$N_1 = 2.7 \; \bar{h} \; (\text{TD})/30 \text{ strokes}/100 \text{ km years} \qquad (4.4)$$

4.2 STROKES TO NEARBY GROUND

The charges in the lower part of the cloud and the leader induce charges of opposite polarity on the ground surface and any conducting elements connected to it. This holds not only for directly grounded objects like ground wires and towers but also for the phase conductors which are tied to ground through transformer neutral connections and remain at ground potential during the relatively slow build-up of the electrostatic field.

A lightning discharge in the vicinity of the line will cause the field to collapse and the charges on the conductors to be released. On any unit length of conductor there appears a voltage equal to charge divided by unit capacity. This static voltage can be represented by two travelling waves of half magnitude propagating in both directions. At the instant of field collapse the two waves are superposed and their equal and opposite currents cancel. Actually the field collapse is not instantaneous; the resultant waves are therefore obtained as the sum of elementary waves generated in each time element. As the elementary waves start their journey successively, the combined wave front will lengthen in proportion to the time taken for the field to collapse, and the wave magnitude will be reduced accordingly.

The induced surges are equal on all three phases, have usually positive polarity, and their wave front is typically 10 μs. Crest values depend on stroke current, distance from stroke, conductor height and the presence or absence of shield wires. As these overvoltages rarely exceed 200 kV, flashovers due to electrostatic induction are unlikely on lines of 33 kV or higher operating voltage[4].

Chowdhuri and Gross[5,6] have, however, drawn attention to the fact that there is also the electromagnetic induction from the return stroke to be considered. Because the inducing current is more or less at right angles to the conductors, they resorted to basic field theory for calculation, and found that the induced voltages can be much higher than those due to the electrostatic effect only. They calculated that the overvoltages can be sufficiently high to flash over the insulation of lines in the higher voltage range. In practice such cases do not seem to be frequent, possibly because strokes that are close enough to induce such high voltages are more likely to hit the line.

4.3 SHIELDING

It has been known for many years that effective shielding requires small shielding angles. A shielding angle (θ_s in *Figure 4.2*) of 30° was considered adequate for lines with towers not higher than 30 m. On lines with portal-type structures, two ground wires are necessary to

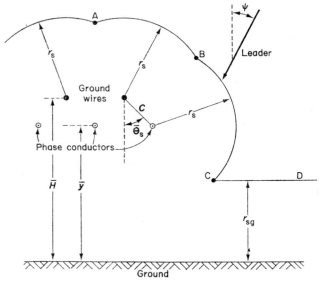

Figure 4.2. Electro-geometric model for shielding analysis (after Brown and White-head[11]):

$\bar{\theta}_s$ = mean shielding angle,
r_s = striking distance,
BC = exposed boundary

meet this requirement. On lines of the 'fir-tree' type, usually double-circuit lines, a moderate increase in shielding angle permitted the use of a single ground wire, with a substantial saving in cost; their performance was satisfactory in areas of low thunderstorm activity.

In the 1950s, e.h.v. double-circuit lines with towers up to 45 m high were constructed in Great Britain and the USA. The outage rates of these lines were considerably higher than had been predicted by methods which had been proven for lower towers. Inadequate shielding was thought to be responsible and this stimulated more thorough study of the shielding problem.

Of the many different methods of assessing shielding performance that have been proposed, two groups stand out: the empirical and the analytical. A useful representative of each will be described. Scale model studies have also been made but are of doubtful validity because the non-linear process of the final breakdown cannot be correctly represented to a smaller scale.

4.3.1 The Burgsdorf-Kostenko Method

In the first group, the work of the Russian engineers is particularly useful as it is based on extensive field records. Burgsdorf[3] published an empirical formula for the probability (P_θ) of shielding failure which expresses the number of strokes by-passing the overhead ground wires and striking a phase conductor as a percentage of the total number of strokes to the line. He assumed that P_θ was a function of shielding angle only but Kostenko et al.[9], after re-examining the field data, proposed equation 4.5 which reveals the influence of the ground wire height (h) at the tower:

$$\log_{10} P_\theta = \theta_s \sqrt{h}/90 - 2 \qquad (4.5)$$

To find the number of shielding failures, one multiplies $P_\theta/100$ by the number (N_l) of strokes to line per 100 km years. It is appropriate to couple P_θ from equation 4.5 with N_l from equation 4.4 as both formulae were derived from the same collection of observations. Only a portion of shielding failures result in flashovers. The critical stroke current magnitude causing flashover is from equation 4.1.

$$I^* = 2V/Z_0 \qquad (4.6)$$

where V is the CFO of the tower insulation and Z_0 the surge impedance of the phase conductor. Burgsdorf also provided an empirical formula for the probability ($P_1\%$) that a stroke will exceed a value (I, kA)

$$\log_{10} P_1 = 2 - I/60 \qquad (4.7)$$

The shielding flashover rate (SFO) is thus given by

$$\text{SFO} = N_l P_1 P_\theta \, 10^{-4} \qquad (4.8)$$

4.3.2 Whitehead's Method

Of the analytical methods that have been developed in the last few years, that of Whitehead et al.[10, 11] is of greatest interest as it has been supported, and substantially verified, by a large scale field investigation, the Pathfinder Project. The analysis proceeds from the idea of the striking distance (r_s), i.e. the distance of the leader head from a ground target at which the average field gradient assumes the critical break-down value. Until the leader head arrives at a striking distance, the point of impact is undecided. When the striking distance is reached, streamers emanate from the ground target to meet the leader. The critical mean gradient has been estimated by various researchers as between 3 and 6 kV/cm. Whitehead allows for the possibility that the mean breakdown gradient to the ground plane may differ from that to the shielding or phase conductors; he accordingly relates the corres-ponding striking distances by a factor (k_{sg}) which may have to be adjusted according to field experience.

$$k_{sg} = r_{sg}/r_s \tag{4.9}$$

The striking distance depends on the charge in the leader which in turn determines the approximate crest value of the prospective stroke current (I_0). Whitehead's relation is

$$r_s = 6.7 I_0^{0.8} \tag{4.10}$$

with I_0 in kA and r_s in metres. For each value of stroke current the striking distances r_s and r_{sg} define a surface ABCD (*Figure 4.2*) of which the portion BC is the exposed boundary, i.e. all strokes crossing BC are assumed to terminate on the phase conductor.

From the geometry of *Figure 4.2* it is clear that the exposed arc BC shrinks with increasing stroke currents, i.e. increasing r_s, until it becomes zero for a striking distance r_{s2}. For smaller currents, hence smaller r_s, the exposure increases but below the critical striking distance (r_{s1}), corresponding to the critical stroke current ($I^* = 2V/Z_0$) flashover cannot occur. For effective shielding

$$r_{s1} \geqslant r_{s2} \tag{4.11}$$

Figure 4.3 gives the critical mean shielding angle θ_{sc} as a function of mean phase conductor height ($\bar{y}$) and mean conductor-to-ground wire spacing ($\bar{c}$), both quantities in per unit of the critical striking distance (r_{s1}). 'Mean' refers to the averages over the span. Field results from the Pathfinder Project are in good agreement with the analysis. In this project, 4600 indicators, capable of distinguishing between direct strokes and backflashovers, and also signalling the occurrence of power follow currents, were installed on 433 miles of transmission lines[10]. Observations have been reported over several years.

Both *Figure 4.3* and the geometrical relationships of *Figure 4.2* indicate that the shielding angle should be reduced with increasing height.

The analysis can be extended to calculate flashover rates for lines having partially effective shielding. For any given striking distance

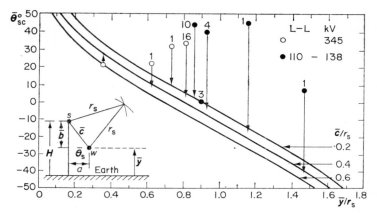

*Figure 4.3. Critical mean shielding angle for effective shielding (for $k_{sg} = 1.0$), in terms of normalised geometry (courtesy Whitehead, discussion IEEE Transactions on Power Apparatus and Systems, **89**, 1908 (1970)). Dots indicate geometry of lines for 38 shielding failure events, arrows indicate recommended shielding angles for same lines.*

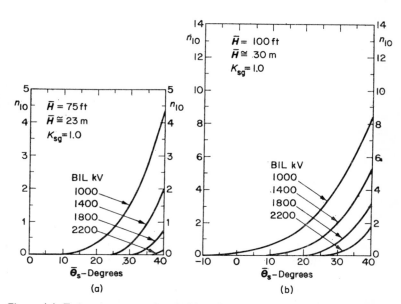

Figure 4.4. Estimating curves for shielding flashover rates, as a function of mean shielding angle $(\bar{\theta}_s)$, mean height of ground wire $(\bar{H})$, and BIL of line insulation. Applicable to groundflash density of $10/mile^2$ year $(= 3.85/km^2$ year). Ordinates in outages/100 mile years $(= /161\ km\ years)$. (Courtesy Whitehead[12].)

or stroke current, the average number of strokes crossing the arc BC in *Figure 4.2* can be determined from geometric considerations. The total number of strokes causing shielding flashovers is obtained by integration between the limit r_{s1} and r_{s2}. As a refinement, Brown and Whitehead assumed a distribution of stroke density as a function of the leader approach angle (ψ in *Figure 4.2*) which is a maximum for vertical strokes and diminishes to zero for horizontal strokes. The precise function used is necessarily arbitrary and subject to empirical verification. In integrating, the geometrical limitations of approach angle have to be taken into account[11].

Curves for estimating shielding flashover rates as a function of line height, insulation level and shielding angle, based on $k_{sg} = 0.9$, were published[11] but in the light of more recent field data from the Pathfinder Project, these curves are superseded by those of[12], calculated for $k_{sg} = 1.0$. Some of these curves are reproduced in *Figure 4.4*.

The shielding criterion, equation 4.11, should be met at every point of the line. Both longitudinal and transverse slopes of the terrain exert an important influence. There is also evidence that trees along the line route have a beneficial effect by raising the effective ground plane. These factors can best be taken into account by a Monte Carlo technique (see 5.3.2) as has been done by Currie *et al.*[13].

Example 4.1   For the 220 kV line of Appendix B, (a) estimate the shielding flashover rate using the methods of Kostenko and Whitehead; (b) redesign the ground wire system, according to Whitehead, to ensure effective shielding, assuming flat terrain.

(a) (i) Kostenko. At the tower $\theta_s = 42.3°$. From equation 4.5, $P_\theta = 3.98\%$. From equation 4.4, $N_1 = 62$. The shielding failure rate is $N_1 P_\theta\ 10^{-2} = 2.47$ failures per 100 km years. Of these only a fraction are due to stroke currents of sufficient magnitude to flashover the line insulation. The CFO of a string of 15 discs, 254×127 mm (10×5 in), which are equivalent to 13 discs 254×146 mm ($10 \times 5\frac{3}{4}$ in), is from *Figure 3.7*: $V = 1200$ kV. The surge impedance of the phase conductor $Z_0 = 480\ \Omega$. The critical stroke current $I^* = 2V/Z_0 = 5$ kA. Using equation 4.7, the probability P_1 of a stroke exceeding this value is 83%. From equation 4.8, the shielding flashover rate is

SFO $= 80\ (3.98)\ 83 \cdot 10^{-4} = 2.05$ flashovers/100 km years.

(ii) Whitehead. Average height of ground wire $\bar{h} = 25.3$ m (83 ft), average shielding angle $\bar{\theta}_s = 37°$.

Interpolating between the curves of *Figure 4.4* for heights of 23 and 30.5 m, the shielding flashover rate for an insulation level of 1200 kV and a groundflash density of 3.86 strokes/km² year ($= 10$ strokes per mile² year, assumed by Whitehead to correspond to TD $= 25$) is

SFO $= 3.4$ flashovers/100 mile years.

Adjusting for TD = 27 and converting to metric units

SFO = 2.3 flashovers/100 km years.

(b) Consider installing two overhead ground wires at the same height as the existing ground wire. The problem is to select a separation distance which will ensure effective shielding. This involves recourse to *Figure 4.3*. Average height of top phase conductor $\bar{y} = 19.2$ m, average vertical separation $\bar{b} = 6$ m.

Critical stroke current from (a): $I^* = 5$ kA. Whitehead suggests a multiplying factor of 1.1 to determine the prospective stroke current (I_0) to a zero-resistance earth (thus accounting for the large but finite surge impedance of the stroke channel compared to the terminating surge impedance). Therefore $I_0 = 5.5$ kA; from equation 4.10: $r_s = 26.3$ m. This is the critical striking distance used for normalising the dimensions $\bar{y}$ and $\bar{c}$. $\bar{y}/r_s = 0.73$. Make a preliminary estimate for $\bar{c}/r_s = 0.2$. From *Figure 4.3*: $\bar{\theta} = 10°$.

It is necessary to check the assumed value for $\bar{c}$.

$\bar{c} = \bar{b}/\cos \bar{\theta}_{sc} = 6.1$ m and $\bar{c}/r_s = 0.23$.

This is sufficiently close to the assumed value. Horizontal separation between the phase conductor and the required location of the ground wire $= \bar{b} \tan 10° = 1.06$ m. The spacing between the overhead ground wires is $9.46 - 2 \times 1.06 = 7.5$ m.

4.4 STROKES TO TOWERS

4.4.1 Basis of calculations

The lightning currents in the stroke channel and towers flow at substantially right angles to those in ground wires and phase conductors. The mutual effects of moving charges in these circuits do not obey conventional travelling wave theory which applies only to parallel conductors. The more accurate method of calculating the voltage stress on the insulation is by field theory concepts. This approach was adopted by Wagner and Hileman[14, 15, 16] and Lundholm et al.[17] but requires very complicated calculations. To adapt it to practical use, Wagner and Hileman[16] derive the voltage across the insulation by the superposition of two voltage components: one due to the current injected into the tower and ground wire system and the other due to the charge above the tower. The effect of the first component preponderates for short wave fronts (up to 1 μs) whilst the influence of the second increases for the longer wave fronts (above 2 μs).

Liew and Darveniza[18] calculated the reduction in the voltage stressing tower insulation caused by a streamer projecting from the tower to meet the opposite polarity leader head. As streamer length is un-known, only an approximate evaluation is possible. Most practical prediction methods consider only the injected-current component and field experience justifies this simplification.

Tower Surge Impedance.   The tower surge impedance has been computed and measured both in the field and on scale models[16, 19, 20]. It is found that it varies along the tower and that considerable atte-nuation exists. It is nevertheless possible to determine a constant value which gives quite closely the same time variation of tower top poten-tial as model tests and field theory calculations.

A recommended formula for tower surge impedance suitable for towers that can be idealised by a cone of height h and base radius r, e.g. conventional double-circuit towers[19], is given by

$$Z_T = 30 \ ln \ 2(1+h^2/r^2) \quad \text{(ohm)} \quad (4.12)$$

For cylindrical towers, the following formula[21] gives satisfactory results:

$$Z_T = 60 \ ln \ h/r + 90 \ r/h - 60 \quad \text{(ohm)} \quad (4.13)$$

where h and r are the height and equivalent radius (mean periphery divided by 2π) of the tower.

Equation 4.13 can be used for downleads, with r denoting the wire radius. For towers which cannot be approximated by cones or cylin-ders, special scale model studies are recommended for the determina-tion of the effective tower surge impedance.

Tower Footing Resistance.   Another important parameter is the resistance in the path of lightning currents entering earth. The power frequency value of grounding electrodes depends on their shape and dimensions and is directly proportional to soil resistivity. The latter ranges from 100 Ωm or less for moist earth to 1000 Ωm for dry earth, and can reach 10^7 Ωm in rocky ground. Calculation and mea-surement of ground resistance are amply covered in the literature. So are measures needed to reduce undesirably high values by means of driven rods and/or long buried conductors (counterpoises) (see Chap-ter 17 of [22]).

Under the influence of high current densities, the soil surrounding ground electrodes becomes ionised or may even suffer partial break-down. As a consequence, the impulse value of tower footing resistance or driven rods is lower than the value measured at power frequency. In the 5–15 Ω range, the effect is unimportant but for high resistances the reduction can be substantial. *Figure 4.5* can be used to estimate the impulse resistance from the power frequency value.

The impulse resistance of long buried counterpoises is a function of time. At the moment of incidence it is equal to 150 Ω, the surge impedance of a buried wire; it then falls to the power frequency value within the time the surge takes to return from the end of the counterpoise. As the velocity is about 100 m/μs, the p.f. resistance is attained in 2 μs per 100 m length. Compared to a single wire of length l, two wires of length $0.5l$, laid in different directions, have an initial resistance of 75 Ω, falling to p.f. value in half the time. A four wire 'crow's foot' is better still.

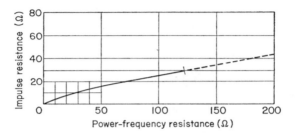

Figure 4.5. Estimating curve for impulse resistance of concentrated tower grounds. Broken line is approximate extrapolation of test values.

4.4.2   Voltage stressing tower insulation

For a stroke to tower, the terminating surge impedance is the resultant of the tower surge impedance Z_T in parallel with half the ground wire surge impedance Z_g (allowing for ground wires in both directions). For two ground wires in each direction, Z_g is to be taken as their equivalent surge impedance (equation A.13 in Appendix A4.2). The towertop potential V_T is obtained from equation 4.2 by substituting for the terminating impedance

$$Z = Z_T/(1+2Z_T/Z_g) \qquad (4.14)$$

V_T is subsequently modified by reflections from the tower base and eventually by reflections from adjacent towers, as discussed in 4.4.4.

When a lightning surge enters a ground wire, the associated current and voltage waves induce in any parallel phase conductor a surge of the same polarity and of K times the ground wire voltage. The value of K is derived in Appendix A4.3. The voltage stressing the tower insulation is the difference between the towertop potential and the voltage induced in the phase conductor

$$V_i = (1-K)V_T = (1-K)I_0Z/(1+Z/Z_s) \qquad (4.15)$$

Since K is of the order of 0.15–0.30 it can be seen that the insulation stress is substantially alleviated by the coupling effect.

The voltages on conductors or ground wires struck by lightning are normally much greater than the corona inception voltage. The corona sheath surrounding the conductor affects the electrostatic field as would an increase in conductor radius. The net effect is an increase in the effective coupling factor and a reduction in the voltage stressing the insulation. Empirical data for effective corona radii of ground wires have been provided by McCann[23].

The greater the distance between coupled conductors, the lower is K. It follows that for a stroke to a tower with vertically arranged conductors, the lower conductors receive less relief than the upper ones but this is compensated by the lower height, as will be shown below.

From the equations of Appendix A4.2 it is seen that two ground wires have a greater coupling to phase conductors than one ground wire; this advantage is additional to the improved shielding mentioned previously. The coupling factor also comes into play when a backflash occurs on a double-circuit line; the affected phase conductor then acts like an additional 'ground wire' with increased coupling to the sound phases and this may obviate a second flashover. A continuous counterpoise also increases the coupling factor.

Example 4.2   Some figures will illustrate the improvement achieved by effective shielding. Assume that a tower of surge impedance 130 Ω is hit by a stroke of current $I_0 = 20$ kA which, according to the AIEE curve of *Figure 2.2*, has a 40% probability of being exceeded. Two parallel ground wires of combined surge impedance $Z_g = 334$ Ω are provided. The terminating surge impedance (Z) is the parallel combination of Z_T and $\frac{1}{2}Z_g$; $Z = 73$ Ω. The surge impedance of the stroke channel is chosen as 1500 Ω; variations from this value have little effect. With $K = 0.30$, equation 4.15 gives for the voltage stressing the insulation

$$V_i = 0.7 \ (20) \ 73/(1+73/1500) = 980 \text{ kV}$$

This figure should be compared to the voltage produced by a direct stroke to a phase conductor of, say, 500 Ω surge impedance:

$$V_i = 20 \ (250)/(1+250/1500) = 4300 \text{ kV}$$

Although corona will reduce this voltage, it is obvious that the stroke current causing flashover will be very much lower than in the case of strokes to the grounding system.

4.4.3   Effect of tower footing resistance

A realistic assessment has to take account of the reflections from the tower base. If the tower footing resistance is much lower than the tower surge impedance—and this is the aim of good design—the wave travelling down the tower is reflected with opposite sign. When the reflected wave arrives at the tower/crossarm junction after a fraction of a microsecond, it is superimposed on the existing tower potential at that point and the rate of rise is thereby substantially reduced. Because of the sloping front of the stroke current, the tower potential has not yet reached its peak value at the instant the reflected wave returns. The longer the wave front is in relation to the return time, the greater is the reduction in magnitude caused by low tower footing resistance.

Example 4.3   Figure 4.6 shows a graphical construction, using a lattice diagram, of the voltage at the tower/crossarm junction for a 1 A current surge with 0.5 μs ramp front and infinite tail. The

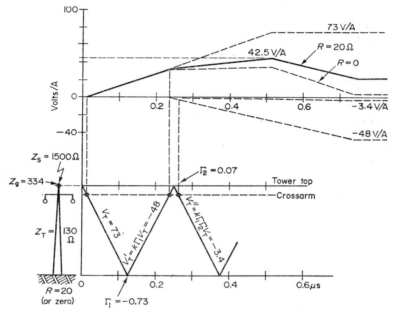

Figure 4.6. Lattice diagram calculation of tower top voltage:
Tower height = 30 m; time to crest = 0.5 μs; tower footing resistance $R = 20\ \Omega$.
Voltage for $R = 0\ \Omega$ also shown.

tower is 30 m high; the wave velocity on the tower is assumed to be 240 m/μs. The resultant voltage has been determined for $R = 0$ and for $R = 20\ \Omega$. For 20 kA, $R = 20\ \Omega$ and $K = 0.7$, the voltage across the insulation is

$$V_i = (0.7)\ 42.5\ (\text{V/A}) \cdot 20\ (\text{kA}) = 595\ \text{kV}$$

(compared to 980 kV calculated without regard to tower footing resistance and sloping front).

5*

It is clear than on towers of moderate height, a low tower footing resistance will be very effective in reducing the tower top potential. The longer the surge front, the more effective will be low footing resistances. In the above example, the reduction is significant despite a rather unrealistically short front time.

On very tall towers, e.g. for river crossings, the potential may well be determined by the terminating surge impedance, i.e. the combination of tower and ground wire surge impedance, rather than by the footing resistance.

A flat-topped ramp function is a convenient approximation to practical lightning waves but typical lightning current oscillograms show concave upwards fronts (see *Figure 2.1*). A repeat of the construction of *Figure 4.6* for this case indicates that the peak voltage is higher. As shown in 3.2.8 the withstand strength of insulators is considerably increased for the short voltage spikes produced by steep-fronted surges.

4.4.4   Effect of adjacent towers

The voltage waves travelling away from the stricken tower on the ground wires return from adjacent towers with opposite sign; if the time to crest of the incident stroke is longer than the return travel time, they will act like the reflected wave from the tower base in further reducing the towertop voltage. If the reflected waves only arrive after the peak of the towertop potential has been reached, they will shorten the tail and still be of some benefit in reducing the integrated voltage stress. A span length of 300 m requires a return travel time of 2 μs which is quite likely to be less than the stroke front.

4.5 STROKES TO GROUND WIRES

4.5.1   Stress at midspan

If the stroke is to a ground wire at some distance from a tower, say at midspan, the voltage on the ground wire is found from equation 4.2 by setting the terminating surge impedance equal to half the self surge impedance of the ground wire: $Z = \frac{1}{2} Z_g$.

$$V_M = \tfrac{1}{2} I_0 Z_g/(1+Z_g/2Z_s) \tag{4.16}$$

The voltage on the nearest phase conductor is KV_M. The voltage $(1-K)V_M$, which the airgap must withstand, is considerably higher than any voltage stress across tower insulator strings that may be caused by a stroke of equal intensity to either tower or ground wire. This follows from the higher effective terminating surge impedance at midspan. Furthermore, the return time for any alleviating reflected wave is much longer at midspan (e.g. 1 μs for a stroke 150 m from the nearest tower) than at a tower position. For these reasons it is usual to increase midspan clearance by sagging the ground wire less than the phase conductors. There is, however, a mitigating circumstance due to the pre-discharge phenomenon described by Wagner and Hileman[24]. As the voltage between ground wire and conductor approaches breakdown value, very large pre-discharge currents begin to flow between them and cause a reduction of the potential difference. This may delay breakdown sufficiently to allow the negative reflections from adjacent towers to arrive in time to prevent the breakdown that would otherwise have occurred. It is indeed an observed fact that midspan flashovers are rare.

4.5.2 Stress to tower insulation after stroke to midspan

Assuming there is no flashover at midspan, the voltage waves V_M and KV_M on ground wire and phase conductor respectively travel towards the adjacent towers where they are modified by reflections. From the equivalent circuits of *Figure 4.1* one finds for the towertop voltage V_T'

$$V_T' = V_M Z_T/(Z_T + \tfrac{1}{2}Z_g) \qquad (4.17)$$

The induced voltage on the phase conductor is reduced in the same proportion; the voltage stressing the insulation is therefore

$$V_i' = (1-K)V_M/(1 + Z_g/2Z_T) \qquad (4.18)$$

As in the case of a stroke to the tower, this voltage is drastically reduced by the arrival of a negative reflected wave from the tower base, provided the tower footing resistance is low.

The maximum stress on tower insulation is of the same order of magnitude as if the tower had been hit directly. This goes to show that strokes to midspan can result in tower flashover though no midspan flashover has occurred.

4.6 PRACTICAL METHODS OF DETERMINING THE VOLTAGES STRESSING LINE INSULATION

In the preceding two sections, discussion was limited to simplified cases; the methods used to deal with the more complex situations encountered in practice will be surveyed in the following. Specific information required for lightning performance predictions concerns the magnitudes and waveshapes of the voltages stressing all critical points of the line insulation. These voltages must be determinable for various combinations of

(1) lightning stroke characteristics: waveshape, current magnitude, point of impact (tower, quarter-span or midspan, ground wire or phase conductor), and
(2) line parameters: span length, tower height and surge impedance, tower footing resistances, self and mutual surge impedance of the conductors.

Clearly many combinations are possible. The method employed must therefore be versatile. Three techniques are commonly used.

4.6.1   TNA or Anacom studies

Several spans of a line are modelled using lumped equivalent impedances for the ground wire(s), the towers and the tower footing resistances. The lightning stroke is replaced by a current generator which is used to inject a small current of appropriate waveshape into the various strike points. The required voltages on the tower/ground wire system are measured oscillographically, usually at the towers and at quarter- and midspan, for a range of span lengths and footing resistances, and for various tower types. As indicated in 4.4.2, the voltage stressing the insulation is $(1 - K)$ times the voltage on the tower–ground wire system (K = coupling factor). The results are usually recorded on a voltage per ampere of stroke current basis.

4.6.2   Scale model studies

Many of the uncertainties in modelling a line by lumped equivalents are avoided by using scale models. An exact model of a section of line, comprising at least three towers, is built to reduced scale (usually 1/25 to 1/50), and is subjected to a stroke current whose time scale must be reduced in the same proportion. The voltages appearing across the line insulation are measured directly, using oscillographs with nanosecond resolution, for various combinations of relevant line parameters and are recorded on a volt/ampere basis.

The model method by-passes many of the difficulties of the purely theoretical approach, e.g. the representation of a complicated tower structure by a simple geometrical shape. Although model studies introduce delicate measurement problems and difficulties in producing a realistic model for the stroke current generator, they have been successfully combined with the mathematical treatment.

4.6.3 Digital methods

Digital computers provide a powerful analytical tool for extending the versatility of Bewley's lattice diagram technique of analysing travelling wave phenomena on lines. In principle, the accuracy of digital computation methods is only limited to the accuracy with which the physical system can be represented. Good agreement is possible between scale model measurements and calculated voltages, provided correct values of surge impedance are ascribed to the line elements, particularly the tower[19]. Computation times on modern high speed computers are such that the digital method can be economically used to determine the voltages stressing line insulations for each combination of a wide range of stroke and line parameters. The method is inherently flexible, as system alterations can be accommodated simply by changing the input data.

Digital methods have an important advantage over the TNA and scale model methods. While the latter are inherently restricted to linear phenomena, the time-iterative procedure of the wave table analysis of travelling waves can take care of non-linearities arising from time-variant and voltage- or current-dependent phenomena. This technique lends itself to the representation of such phenomena as corona attenuation and distortion, corona effects on self and mutual impedances, current-dependent footing resistances, and voltage-dependent flashover phenomena. Such capability is essential for predicting the likelihood of multiple flashovers, as required for example in determining the double-circuit outage rate of overhead lines (see 5.3.4).

4.7 ATTENUATION AND DISTORTION OF LIGHTNING SURGES

For surges travelling short distances, as in the applications so far considered, the assumptions of unchanging waveshape and exponential decay of crest are adequate. A more accurate description of attenuation and distortion is needed for surges travelling over several spans. This situation arises in the insulation co-ordination of substations (Chapter 7), when it is important to take advantage of the significant reductions in front slope and crest incurred by surges originating 1–3 km from the substation.

4.7.1 Corona effects

Attenuation and distortion are caused by energy losses and corona effects. For steep wave fronts, skin effect in conductors and ground return paths produces some attenuation and distortion but by far the most drastic reduction of lightning surges is caused by corona. As the surge voltages are always much greater than the corona inception voltage (V_c), the corona discharge not only adds to the energy losses but has also the effect of retarding the portion of the wave front above corona inception voltage, resulting in the characteristic shearing back of the wave depicted in *Figure 4.7a*. The slowing down of that part

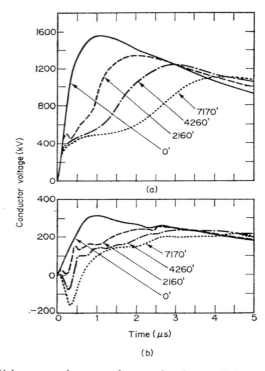

Figure 4.7. Voltage waveshapes on the surged and a parallel unsurged conductor (9.7 m flat spacing) at various distances (marked in feet) from the origin. Positive polarity surge applied to a 50.8 mm (2.0 in) diameter, steel-cored aluminium conductor. (Courtesy Wagner et al.[26].)

of the wave is explained by the increase in effective capacitance associated with the radially conducting corona sheath.

Skilling and Dykes[25] have shown that this corona effect can be treated mathematically by assigning an appropriate propagation velocity to each voltage value (V) on the wave front which exceeds V_c. From this assumption they derive the relationship between retardation

(Δt) and distance travelled (D):

$$\Delta t/D = (k/uC)(1 - V_c/V) \qquad (4.19)$$

where

k = empirical constant determined by field tests
C = capacitance of conductor, in $\mu F/m$
u = propagation velocity for $V < V_c$, in $m/\mu s$.

The corona inception voltage in kV for single or bundle conductors is

$$V_c = 1.66 \cdot 10^{-3} nmk_d^{2/3} r(1 + 0.3/\sqrt{r})/k_b C \qquad (4.20)$$

where

n = number of sub-conductors in bundle
m = conductor surface factor (normally 0.7–0.8)
k_d = relative air density
r = conductor radius, in cm
C = capacitance as above
k_b = ratio of maximum to average surface gradient for bundle conductors

$$k_b = 1 + 2(n - 1) \sin (\pi/n)(r/A) \qquad (4.21)$$

where A is the distance between adjacent sub-conductors, in cm.

The capacitance (C) can be found from reference tables or from the slightly simplified formula

$$C = 10^{-3}/41.6 \log (2h/r_e) \qquad (4.22)$$

where

h = average height above ground, in cm
r_e = effective radius of bundle, in cm
$r_e = rn^{1/n} [A/(2r \sin \pi/n)]^{(n-1)/n}$ \qquad (4.23)

It will be seen that for $n = 1$, $r_e = r$ and $k_b = 1$.

4.7.2 Distortion caused by mode-propagation

A further distorting effect exists on multi-conductor lines. It can be analysed by resolving the voltages into line components and ground components, each set having different velocities of propagation and different rates of attenuation. For surges below the corona inception voltage, the velocity of propagation of the line component is nearly the velocity of light and distortion is negligible. For ground components, the electrostatic charges induced in the ground are near the surface while the return current is well below the surface, at a depth dependent on frequency and earth resistivity. For a perfectly conducting earth, the return current would flow at the ground surface and the velocity would be equal to the velocity of light; in reality, the effective distance of the conductor from its image is increased and the velocity correspondingly reduced. The mode propagation distortion results then from the different velocities of the surge components.

Consider a three-conductor line carrying surge voltages to ground (V_1, V_2, V_3). The ground component v_e is defined by

$$v_e = \tfrac{1}{3}(V_1 + V_2 + V_3) \tag{4.24}$$

and the line components as

$$v_1 = V_1 - v_e; \quad v_2 = V_2 - v_e; \quad v_3 = V_3 - v_e \tag{4.25}$$

By substitution of equation 4.25 in equation 4.24 it can be seen that the sum of the line components is zero, hence there are no ground currents associated with the line components; this explains the small attenuation.

From equation 4.24 it follows that for equal surges on all three phases, there exists only a ground component. For a single-phase surge (V_1), accompanied by induced surges (KV_1) on the other two phases $(K = \text{coupling factor})$, the ground component is from equation 4.24: $v_e = \tfrac{1}{3}(1+2k)V_1$; the line components are from equation 4.25: $\tfrac{2}{3}(1-K)V_1$ on the one phase and $-\tfrac{1}{3}(1-K)V_1$ on the other two, i.e. the charge on phase 1 is balanced by charges of opposite polarity of equal total value on phases 2 and 3. *Figure 4.8* illustrates schematically the resolution into components, the distortion produced by their differences in velocity, and the different attenuation of the two components after they have travelled some distance. This effect contri-

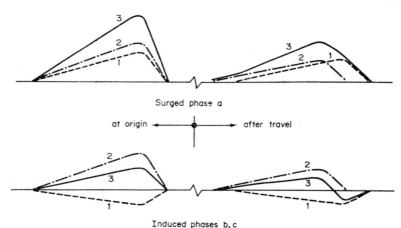

Surged phase a

at origin ← ⟶ after travel

Induced phases b, c

Figure 4.8. Mode propagation and distortion on phase-balanced three-phase line:
1, line components,
2, ground components,
3, total voltage wave

buted to the distortion of the wave shown in *Figure 4.7a* and *Figure 4.7b* displays the typical negative dip of the wave induced in a coupled conductor.

4.7.3 Simple estimating formulae

From numerous oscillograms similar to *Figure 4.7*, Wagner *et al.*[26] and others obtained information on the overall distortion suffered by

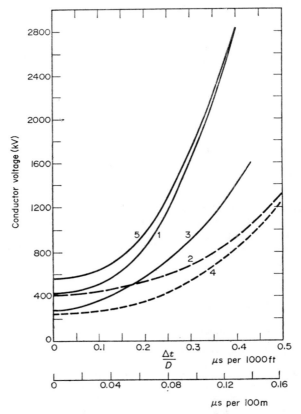

Figure 4.9. Slope reduction of travelling waves:

1, 2,   steel-cored aluminium conductor, 50.8 mm (2.0 in) diameter;
3, 4,   SCA conductor 23.6 mm (0.927 in) diameter;
5,   SCA twin conductors, 42.7 mm (1.68 in) diameter,
          bundle spacing 456 mm (18 in);
1, 3, 5, negative polarity surges;
2, 4,   positive polarity surges.
(Curve 5 after Hileman *et al.* IEEE Transactions on Power Apparatus and Systems, **86,** 659 (1967); other curves after Wagner *et al.*[26].)

single-pole surges above the corona level. *Figure 4.9* gives data on the retardation of the wave front which obviate the calculations outlined in connection with equation 4.19.

The intersection of the retarded wave front with the tail of the surge gives an approximation of the crest reduction. A search of the literature will usually yield information from field tests applicable to a particular line (see Chapters 15 and 16 of[22], also[25, 27] for a bibliography on the subject). Failing this, the following can serve as a rough guide:

Front time prolongation:

  below V_c: 0.3 μs/km
  above V_c: 0.6 μs/km for negative surges
                1.2 μs/km for positive surges.

Crest attenuation:

  below V_c: 3%/km for the first 5 km; 1.2–2%/km for the next 30 km
  above V_c: use the Foust and Menger[22] formula:

$$V(D) = V_0/(k'DV_0 + 1) \qquad (4.26)$$

where V_0 and $V(D)$ are the crest voltages at the origin and after a travel of D km, respectively; and k' is a constant which increases from 0.0001 for waves of about 20 μs half time to 0.0002 for waves of about 5 μs half time. For chopped waves, k' can assume even higher values of up to 0.000 45.

REFERENCES—CHAPTER 4

1. Golde, R. H., 'The Frequency of Occurrence and the Distribution of Lightning Flashes to Transmission Lines', *AIEE Transactions Pt. III*, **64**, 902–910 (1945).
2. AIEE Committee Report, 'A Method for Estimating Lightning Performance of Transmission Lines', *AIEE Transactions Pt. III*, **69**, 1187–1196 (1950).
3. Burgsdorf, V. V., 'Lightning Protection of Overhead Transmission Lines and Operating Experience in the U.S.S.R.', *Proceedings CIGRE*, Report 326 (1958).
4. Golde, R. H., 'Lightning Surges on Overhead Distribution Lines Caused by Indirect and Direct Lightning Strokes', *AIEE Transactions Pt. III*, **73**, 437 (1954).
5. Chowdhuri, P., and Gross, E. T. B., 'Voltage Surges Induced by Lightning Strokes', *Proceedings IEE*, **114**, 1899–1906 (1967).
6. Chowdhuri, P., and Gross, E. T. B., 'Voltages Induced on Overhead Multiconductor Lines by Lightning Strokes', *Proceedings IEE*, **116**, 561–565 (1969).
7. Willheim, R., and Waters, M., *Neutral Grounding in High-voltage Transmission*, Elsevier, New York, London, (1956).
8. Peterson, H. A., Rudge, W. J., Monteith, A. C., and Ludwig, L. R., 'Protector Tubes for Power Systems', *AIEE Transactions*, **59**, 282–288 (1940).

9. Kostenko, M. V., Polovoy, I. F., and Rozenfeld, A. N., 'The Role of Lightning Strokes to the Conductors bypassing the Ground Wires in the Protection of H. V. Lines', in Russian—*Elektrichestvo*, **No. 4**, 325–335 (1961).
10. Armstrong, H. R., and Whitehead, E. R., 'Field and Analytical Studies of Transmission Line Shielding', *IEEE Transactions, Power Apparatus and Systems*, **87**, 270–271 (1968).
11. Brown, G. W., and Whitehead, E. R., 'Field and Analytical Studies of Transmission Line Shielding—II', *IEEE Transactions, Power Apparatus and Systems*, **88**, 617–626 (1969).
12. Whitehead, E. R., *Final Report to the Edison Electric Institute, Research Project No. 50*, (to be published).
13. Currie, J. R., Liew, A. C., and Darveniza, M., 'Monte Carlo Determination of the Frequency of Lightning Strokes and Shielding Failures on Transmission Lines', *IEEE Transactions, Power Apparatus and Systems*, **90**, No. 5, 2305 (1971).
14. Wagner, C. F., 'A New Approach to the Calculation of the Lightning Performance of Transmission Lines', *AIEE Transactions Pt. III*, **75**, 1233–1256 (1956).
15. Wagner, C. F., and Hileman, A. R., 'A New Approach to the Calculation of the Lightning Performance of Transmission Lines—Part II', *AIEE Transactions Pt. III*, **78**, 996–1081 (1959).
16. Wagner, C. F., and Hileman, A. R., 'A New Approach to the Calculation of the Lightning Performance of Transmission Lines—Part III', *AIEE Transactions Pt. III*, **79**, 589–603 (1960).
17. Lundholm, R., Finn, R. B., and Price, W. S., 'Calculation of Transmission Line Lightning Voltages by Field Concepts', *AIEE Transactions Pt. III*, **76**, 1271–1283 (1968).
18. Liew, A. C., and Darveniza, M., 'A Sensitivity Analysis of Lightning Performance Calculations for Transmission Lines', *IEEE Transactions, Power Apparatus and Systems*, **90**, 1443–1451 (1971).
19. Sargent, M. A., and Darveniza, M., 'Tower Surge Impedance', *IEEE Transactions, Power Apparatus and Systems*, **88**, No. 5, 680–687 (1969).
20. Kawai, M., 'Studies of Surge Response on a Transmission Tower', *IEEE Transactions, Power Apparatus and Systems*, **83**, 30–34 (1964).
21. Jordan, C. A., 'Lightning Computations for Transmission Lines with Overhead Ground Wires', *General Electric Review*, **37**, No. 3, 130–137 (1934).
22. Westinghouse Electric Corp., *Transmission and Distribution Reference Book*, 4th edn., East Pittsburgh, Pa., (1950).
23. McCann, G. D., 'The Effect of Corona on Coupling Factors between Ground Wires and Phase Conductors', *AIEE Transactions*, **62**, 818–826 (1943).
24. Wagner, C. F., and Hileman, A. R., 'Effect of Pre-discharge Currents on Line Performance', *IEEE Transactions, Power Apparatus and Systems*, **82**, 117–131 (1963).
25. Skilling, H. H., and Dykes, P. de K., 'Distortion of Travelling Waves by Corona', *AIEE Transactions*, **56**, 850–857 (1937).
26. Wagner, C. F., Gross, I. W., and Lloyd, B. L., 'High-voltage Impulse Tests on Transmission Lines', *AIEE Transactions Pt. III*, **73**, 196–210 (1954).
27. Hylten-Cavallius, N., and Annestrand, S., 'Distortion of Travelling Waves in Power Cables and on Power Lines', *Proceedings CIGRE*, Report 325, Appendix III (1962).

5

The Lightning Performance of Transmission Lines

5.1 INTRODUCTION

To design lines of an acceptable lightning performance the transmission engineer must have at his disposal reliable methods of predicting their flashover and outage rates. The available methods, which have been developed on the basis of the techniques described in Chapter 4, can be divided into those making use of generalised curves and those applying a Monte Carlo technique. The former are convenient for quick estimation but are necessarily restricted in the range of line and lightning parameters and give moderate accuracy. The latter provide instructions for the application of probabilistic techniques to the detailed analysis of specific lines for any desired lightning parameters and yield results of better accuracy.

Since all methods of analysis rely on incomplete basic knowledge, they have to be aligned with field experience. Concepts and assumed data giving results in accord with experience are retained, others are rejected. Because of the large number of parameters, this procedure does not eliminate the possibility of compensating errors which may show up when a method is applied to a new type of line and conditions different from those previously encountered. It is therefore a good rule, when applying a prediction method to a new line, to check it whenever possible against operating experience on an existing line, preferably in the same area. It is also undesirable that the user should modify any particular assumption of a method without good reason, for in so doing he may discount the experimental foundation on which it rests.

All prediction methods aim at determining

the frequency (N_1) with which lightning will strike a line (per 100 km year),

the proportion (p_1) of strokes which will cause insulation flashover,
the proportion (p_2) of flashovers which will cause power faults.

From this information are derived

the flashover rate $(FO) = N_1 p_1$ flashovers/100 km years $\qquad$ (5.1)

the outage rate $\quad (OR) = (FO)p_2$ outages/100 km years $\qquad$ (5.2)

5.2 THE FLASHOVER RATE OF UNSHIELDED LINES

As shown in 4.1, a direct stroke to a phase conductor nearly always causes flashover of line insulation. Flashovers may involve phase-to-phase insulation (V_{pp}) as well as phase-to-ground insulation (V_{pg}). The critical lightning currents are given by

$$I_1^* = 2V_{pg}/Z_0 \qquad (5.3a)$$

and

$$I_2^* = 2V_{pp}/Z_0(1-K) \qquad (5.3b)$$

For unbonded and unearthed wood pole lines with high phase-to-ground insulation strength, I_2^* is normally less than I_1^*. Interphase flashovers are therefore to be expected. The probability (p_1) that a stroke current will exceed the smaller I^* may be determined from *Figure 2.2*, and N_1 from equation 4.3 or 4.4. The flashover rate is then found from equation 5.1.

Unshielded lines are sometimes characterised by a variety of structure types, some of which may be 'weak-links' having low insulation strengths. The prediction of lightning performance for such lines can be complex[1] because the weak-link structures can have a disproportionate influence on the outage rate. If they cannot be eliminated, surge protective devices should be installed nearby.

Example 5.1 Estimate the flashover rate of the 220 kV line described in Appendix B, but operated without a ground wire.

The average height of the top conductor
$$= 25.3 - 2/3 \ (9.1) = 19.2 \text{ m.}$$

From equation 4.4, $N_1 = 47$.

The critical flashover voltage for 15 discs is 1200 kV,

i.e. $V_{pg} = 1200$ kV.

The flashover voltage of the phase-to-phase insulation, (i.e. a 5.2 m (17 ft) air gap or two insulator strings in series) is obviously much greater than 1200 kV, and so in this case $I_1^* < I_2^*$.

$$I_1^* = 2 \times 1200/480 = 5 \text{ kA.}$$

From *Figure 2.2*, Curve 1, $p_1 = 0.9$.

$$FO = 47 \times 0.9 = 42 \text{ flashovers}/100 \text{ km years.}$$

5.3 THE FLASHOVER RATE OF SHIELDED LINES

Faults on shielded lines may be caused by flashovers due to shielding failure or by backflashovers. Methods for determining the shielding flashover rate (SFO) have been considered in 4.3. Predictions of back-flashover rates (BFO) may be made using either generalised-curve methods or Monte Carlo analyses.

5.3.1 Prediction of backflashover rates using generalised curves

(1) *The AIEE Method*

A widely used method is that prepared by an AIEE Committee[2] in 1950 on the basis of previous work by Harder and Clayton[3]. It has proved reasonably successful for lines operating at 230 kV and below (though it must be noted that the shielding configurations recommended are now known to be unsuitable for tall towers, see 4.3).

In the Anacom studies used by the AIEE Committee to determine the voltage waveshapes stressing line insulations, a standard tower configuration (height 100 ft, average ground wire height 80 ft) was represented by a 20 μH inductance. A stroke current waveshape of 4/40 μs was selected as being representative of crest currents in excess of 30–40 kA (which a well designed line ought to be able to withstand). The Anacom studies were performed for a wide range of tower footing resistances and spans; strike points investigated were restricted to tower and midspan, and equal probability was assigned to each. For each combination, the recorded voltage waveshapes were analysed and the critical stroke current I^* which could just flashover line insulation was computed from

$$I^* = CV/V_T(1-K) \qquad (5.4)$$

where V = insulation flashover voltage at a specified reference time lag, corrected for atmospheric conditions;

C = a correction factor accounting for the non-standard waveshape of V_T;

V_T = tower top or midspan potential, as the case may be, in volts/ampere of stroke current.

The critical currents for a range of line insulations are presented in generalised curves, e.g. *Figures 5.1* and *5.2*. The flashover rate for each combination is then determined using equation 5.1; p_1 is determined from *Figure 2.2* (Curve 1) and N_1 from equation 4.3. Flashover rates are determined separately for strokes to tower and to midspan, and are added together with suitable weighting factors; usually the mean of the tower and midspan flashover rates is taken except in

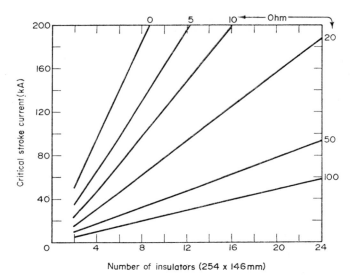

Figure 5.1. Curves for estimating critical stroke current to tower v. number of standard suspension insulators (254 × 146 mm) with tower footing resistance (impulse value) as parameter: span 366 m (1200 ft); thunderday level 30 (after AIEE Committee[2])

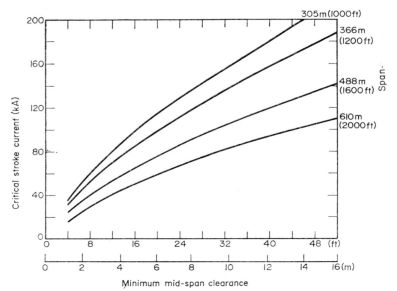

Figure 5.2. Curves for estimating critical stroke current to midspan v. midspan clearance, with span as parameter: thunderday level 30 (after AIEE Committee[2])

6

those circumstances when midspan strokes cause flashovers at the tower rather than across the midspan air gap.

The generalised curves were prepared for the standard tower configuration. In practice, the tower and conductor configuration may vary from that of the standard tower, thus changing the effective value of coupling factor (K) used in equation 5.4. This variation is accounted for by applying a correction factor (*Figure 5.3*) to the number of insulators with which to enter the curves.

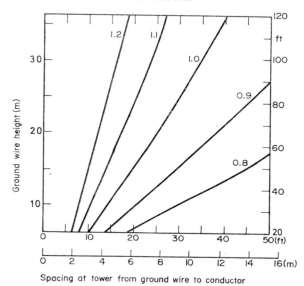

Figure 5.3. Curves for adjusting number of insulators in Figure 5.1 when height and spacing of conductors differ from reference value. Multiply actual number of insulators by correction factor. (After AIEE Committee[2].)

A step-by-step procedure for determining the lightning performance of a line according to the AIEE method is illustrated in Example 5.2.

The AIEE method failed to predict the high flashover rates experienced on the first 345 kV double-circuit lines in the USA which had towers considerably higher (45 m) than those normally used until then, and which were shielded with only one ground wire. As already discussed in 4.3, one of the suspected causes was inadequate shielding, and this is still considered correct. In the search for other explanations, certain of the simplifying assumptions of the AIEE method were held responsible for the low predictions, in particular the representation of the tower as a lumped element and the exclusive use of the 4 μs wave front. These doubts led to the development of more refined generalised curve prediction methods and to the Monte Carlo based methods.

Example 5.2   Estimate the backflashover rate of the 220 kV line described in Appendix B using the AIEE method[2].

Step 1. Determination of the effective number of insulators which represent the insulation strength of the line:

The insulation level of 15 discs, 254×127 mm (10×5 in) equals that of 13 standard discs, 254×146 mm ($10 \times 5\frac{3}{4}$ in)

Ground wire height (at tower) = 30.5 m (100 ft)

Distance to furthest phase conductor = 16.2 m (53.3 ft)

Coupling factor correction = 0.9 (*Figure 5.3*)

Effective number of discs = 12.

Step 2. Distribution of surge-reduced tower footing resistances:

The measured power-frequency resistance values are converted to surge-reduced values with the help of *Figure 4.5*, and the number of bands are contracted to four.

Band identification	1	2	3	4
Surge-reduced resistance,	5	10	15	21
per cent frequency	44	16	31	9
Critical tower stroke current (I_T^*) kA	195	148	115	90
Tower flashover rate (TFR)	0.07	0.22	0.50	1.20

Step 3. Determination of critical tower stroke currents I_T^* and tower flashover rates, TFR:

The critical tower stroke currents for each resistance band are obtained from *Figure 5.1*; the corresponding tower flashover rates are determined using *Figure 2.2* and equation 4.3; these are also entered in the table.

Step 4. Determination of critical midspan stroke currents (I_M^*), and flashover rates, MFR:

The midspan clearance between the earthwire and the nearest phase conductor = 8.25 m (27 ft). The midspan critical stroke current $I_M^* = 122$ kA from *Figure 5.2*.

$$\text{MFR} = 0.45.$$

Step 5. Determination of line backflashover rate:

The backflashover rate for the line is the weighted average of the flashover rate for each band, whence

$$\text{BFO} = \tfrac{1}{2}(0.44 \times 0.07 + 0.16 \times 0.22 + 0.31 \times 0.5 + 0.09 \times 1.2) + \tfrac{1}{2} \times 0.45 = 0.39.$$

Step 6. Adjusting for a thunderday level of 27
BFO = 0.36 backflashovers/100 km years.

(2)   The Clayton and Young Method

In 1964, Clayton and Young[5] introduced a number of refinements into the AIEE method, with the aim of improving prediction for e.h.v. lines. They prepared estimating curves for double-circuit towers of vertical configuration for operating voltages from 115 to 345 kV and for single-circuit towers of horizontal configuration for 115–700 kV, both types with two ground wires, in recognition of the need for effective shielding. The variables are appropriate ranges of spans, number of insulators and tower footing resistances. Guidance is also given on the selection of counterpoises for reducing tower footing resistances. In their calculations and analogue computer work, they assumed lightning stroke front times of 2, 4, and 6 μs, and varied the proportion of strokes with these front times according to stroke current magnitudes, so as to reflect the observed trend to longer front times for the higher current crests. The time to half value was assumed to be 40 μs throughout. Account was taken of the additional stress caused by the instantaneous power frequency voltages which, for e.h.v. lines, is no longer negligible in comparison with the impulse voltage. An allowance for adverse weather was made in the flashover values of the insulator strings.

(3)   The General Electric-Edison Electric Institute Method

In 1968, Anderson, Fisher and Magnusson[6] published generalised curves for estimating the lightning performance of e.h.v. lines. The developed method is a generalisation of detailed analyses made on specific 345, 500 and 735 kV line configurations selected as base cases. It makes use of scale model tests[7] and Monte Carlo techniques. The results are presented in the form of estimating curves (Chapter 8.3 of[6]) for standard tower configurations. Correction factors must be applied if the height of the design case tower differs significantly from that of the base case.

5.3.2 The Monte Carlo Method of calculating backflashover rates

The parameters which determine the lightning performance of a transmission line are subject to large variations according to frequency distribution laws which can only be determined by field observations. The most important are the current magnitudes of strokes to lines, the times to crest, the weather conditions which influence the insulation strength (also indirectly by the swing angle), the lower footing resistances, the strike locations, the chances of shielding failures and the values of power frequency voltages at the instant of stroke. There is little statistical correlation between these quantities.

The Monte Carlo method is based on random sampling of variables which influence the outcome of any lightning incident. Where there is a correlation between parameters, as for example between stroke magnitude and time to crest, it is taken into account.

The randomly selected parameters for each lightning incident are utilised in a calculation to determine the outcome of the incident, namely whether or not the voltage is sufficient to produce a flashover of line insulation. This analysis must be repeated for many hundreds of such incidents to obtain a realistic estimate of the long-term average flashover rate. The number of incidents normally corresponds to the total number of strokes hitting the line over the required period of study, usually 10–20 years.

Anderson[8] was the first to apply the Monte Carlo technique to the prediction of line performance. He demonstrated that a markedly better agreement between predicted and observed line performance was possible with the Monte Carlo method than with the generalised methods, provided realistic input data were selected for the program. Typically, the ratio of observed to calculated outage rates varied between 0.5 and 1.5. An advantage of the Monte Carlo method is that provision can be made in the program for amending or adding new field data as they become available.

Anderson and Barthold[9] refined the statistical approach by a computer program dubbed 'Metifor' (meteorologically integrated forecasting) which creates a digital model of the climate along the proposed transmission line route. It is based on the past *hourly* records of the nearest first-order weather station, covering precipitation rate and type, thunderstorm activity, air density, humidity, wind direction and strength. From these are deduced not only the actual statistics and their variability in time but also the correlations between them as they occurred in nature for the particular location. The program computes hourly 'degradation factors' for the insulation and compares the resulting insulation strength with the insulation stresses caused by lightning, switching surges and power frequency voltage, taking into account their incidence and variability. In this way the low probability of coincidence of severe conditions is properly brought into the calculation, which can readily be extended over a 10–20 year period to cover long-term weather cycles.

5.3.3 Wood pole lines

It is well known that wood can add considerably to the impulse insulation strength of porcelain strings. Lusignan and Miller[10] conducted a large number of laboratory tests on the impulse strength of various wood/porcelain combinations. Their results are represented in a set of curves, *Figure 5.4*, which, in conjunction with *Figure 3.7*, facilitate the determination of the strength of the combination at various time lags, expressed as an equivalent number of standard insulators.

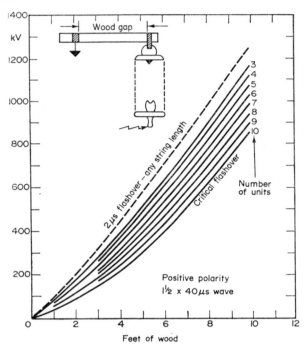

5.4. Impulse insulation added to insulator string by dry wood. Equivalent insulators to be added may be determined from Figure 3.7. (After AIEE Committee².)

More recently studies of the impulse voltage distribution between wood and porcelain[11] produced the following approximate formula for the impulse strength (V_t) of a wood/porcelain combination:

$$V_t = (V_p^2 + V_w^2)^{1/2} \tag{5.5}$$

where V_p is the flashover voltage of the porcelain alone and V_w that of the wood. For wet wood, V_w can be taken as 300 kV/m at minimum flashover level, and as 1.6 and 2.0 times this value at time lags of 4 and 2 μs respectively. Comparisons between measured and estimated flashover voltages indicate that the insulation strength of wood/porcelain combinations can be estimated to an accuracy of ±15%. To extract the greatest benefit from the insulating property of wood, parallel air and wood/porcelain flashover paths on a pole top must be carefully co-ordinated (see 5.4.2). This requires better accuracy which can only be obtained by laboratory proving tests.

Example 5.3 Estimate the insulation strength (in equivalent insulators) of the 220 kV line of Appendix B, but with the steel crossarms replaced by insulating arms constructed of wood; effective length of wood in arm is 3 m.

(a) From *Figure 5.4*, the additional strength at 2 μs provided by 3 m of dry wood is 1230 kV. When added to the 1630 kV 2 μs strength of 13 standard discs, this gives a combined strength of 2860 kV, which is equivalent to 23 discs (from *Figure 3.6*). Thus, the dry wood has an equivalent additional strength of 10 discs. The relevant impulse strength is, however, associated with wet wood. For this the AIEE Committee recommends that only half the additional number of discs be relied on. Therefore, the equivalent strength is 18 standard discs.
(b) At twice 300 kV/m (for a 2 μs time lag), the impulse strength of 3 m of wet wood alone is 1800 kV. The 2 μs strength of 13 discs is 1630 kV. Therefore, from equation 5.5, the combined strength is 2430 kV; this is equivalent to 20 standard discs (from *Figure 3.6*).

5.3.4 Double-circuit lines

Because of scarcity of right-of-ways, double-circuit lines form an increasing proportion of lines built. A significant percentage of lightning faults on double-circuit lines can affect both circuits; the field data range between 0 and 70%, with an average value of about 30%. In many cases, critical system conditions result and it is important to be able to predict the double-circuit outage rate and, if possible, reduce it by appropriate design.

Sargent and Darveniza applied the Monte Carlo technique of 5.3.2 to the calculation of double-circuit outages[12] and succeeded in achieving good agreement with outage statistics, including such details as the proportion of double-circuit outages and fault types[13]. Some actual lightning incidents on a 220 kV line, for which the stroke currents and affected phases had been recorded, could be correctly simulated. The computer program takes account of the changes in surge impedances, reflection coefficients and coupling factors caused by the first and second flashover, if it occurs, and also models the voltage-dependent corona effects.

5.4 OUTAGE RATE AND SUSTAINED OUTAGE RATE

Lightning flashovers do not always lead to power follow current and consequent circuit interruptions. Moreover, service experience indicates that few lightning flashovers damage the line insulation permanently. The number of sustained interruptions can therefore be drastically reduced by reclosing the line after a short period of de-energisation during which the fault arc is likely to go out spontaneously. High-speed protection is desirable to keep fault damage down. With automatic reclosing the success rate is of the order of 90%.

As a rough indication of the outage rates normally expected and considered acceptable, the following figures are quoted:

11–66 kV rural lines	7–3 outages/100 km years
132 kV lines, in general	0.6 outages/100 km years
Very important 132 kV and extra high voltage lines	0–0.3 outages/100 km years

Sustained outage rates are about one tenth of the above figures where automatic reclosing is employed.

5.4.1   Steel tower lines

For steel tower lines with porcelain insulation only, the probability (p_1) of a flashover developing into an outage is approximately 0.85. On shielded transmission lines in areas of high isokeraunic level, the use of automatic reclosing produces a substantial improvement in performance. In areas of low thunderstorm activity, line costs can be cut by omitting ground wires and avoiding costly work for the reduction of high footing resistances. Reliance is then placed on automatic reclosing to obtain adequate reliability[16].

Example 5.4   The Monte Carlo program of Sargent and Darveniza referred to in 5.3.4 was applied to the double-circuit line of Appendix B, simulating 10 years' operation of a 100 mile (161 km) section. The number of flashovers of different types were as follows:

shielding flashovers	16
backflashovers on one circuit	2
backflashovers on both circuits	6
Total	24

Assuming an outage probability $p_2 = 0.85$, the total outage rate becomes 1.27 outages/100 km years, with 25% double-circuit outages. In actual operating experience of 1500 km years, a total outage rate of 1.02/100 km years, with 33% double-circuit outages, was recorded. The agreement is within the accuracy that can be expected from such calculations.

Summating the results of Example 4.1 for shielding failures and Example 5.2 for the backflashover rate, and multiplying by $p_2 = 0.85$, one arrives at an estimate of 2.1–2.3 outages/100 km years. The methods used in the latter examples do not permit an estimate of double-circuit flashovers.

5.4.2 Wood pole lines

For shielded lines which are so designed that all flashovers take place across the air clearance to bonding or downleads, p_2 is about 0.85 as for steel tower lines. In this case, the likelihood of arc damage to the insulation is small and the success rate for automatic reclosing high.

If, on the other hand, the arc quenching property of wood is to be exploited, the insulation must be so co-ordinated that the preferred discharge path is via the wood/porcelain insulation. Field observations and laboratory experiments have established that to maintain an arc in contact with wood it is necessary that a certain threshold voltage gradient is exceeded. At a low power-frequency voltage gradient along the wood path the arc is likely to go out. The probability (p_2) of the impulse discharge being continued by a power arc is, according to some investigators[1, 14, 15] given by the curves of *Figure 5.5*. For a voltage

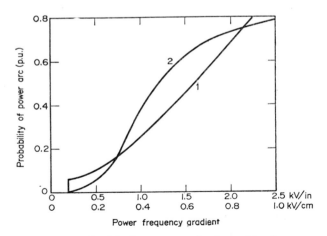

Figure 5.5. Probability (p_2) of flashover on wood pole resulting in a power arc, as a function of power frequency voltage gradient across wood insulation: 1, after Ekvall[15], 2, after Darveniza et al.[1]

gradient of 0.2 kV/cm, p_2 is only 5–10%. By appropriate design it is possible to suppress about 90% of outages.

Unshielded lines have a very high outage rate (see Example 5.1) but by taking advantage of the arc quenching effect of wood, reinforced by automatic reclosing, even cheaply constructed lines can achieve good reliability, comparable with that of shielded lines.

A reservation has to be made in respect of lines which are so designed that they are likely to suffer interphase flashovers (see 5.2). For these lines a much less favourable performance must be expected.

If structures are designed to utilise the arc quenching property of wood, they are in danger of the lightning current shattering the pole or crossarm. The impulse arc penetrates into unseasoned wood and laboratory tests show that surge currents of less than 10 kA$_p$ can shatter a crossarm or pole[1]. In well seasoned timber, on the other hand, service experience and laboratory tests show that the arc hugs the surface and the only damage is usually splinters of wood blown off. Pressure impregnated poles seem to retain moisture for a longer time and are therefore more susceptible to damage. Field data indicate that, apart from newly constructed lines, the incidence of permanent damage on wood pole lines is low.

Example 5.5   Estimate the outage rate of the 220 kV line of Appendix B but without a shielding wire and with wood crossarms.

From Example 5.4, the wood-porcelain insulation is equivalent to 18–20 standard discs; select 18 with a CFO of 1600 kV. Then proceeding as in Example 5.1,

$$FO = 38.5 \text{ flashovers}/100 \text{ km years.}$$

Because phase-to-phase insulation is high, and assuming reasonably low footing resistances for the towers, most incidents will only involve single-phase flashovers. Curve 2 of *Figure 5.5* should be used to determine the outage/flashover ratio. The operating gradient is $220 \text{ kV}/(\sqrt{3}.300 \text{ cm}) = 0.42 \text{ kV r.m.s.}/\text{cm}$ of wood, and this gives $p_2 = 0.4$. Then $OR = 38.5 \times 0.4 = 15.5$ outages/100 km years. With reclosing about two sustained outages could be expected per 100 km years.

REFERENCES—CHAPTER 5

1. Darveniza, M., Limbourn, G. J., and Prentice, S. A., 'Line Design and the Electrical Properties of Wood', *IEEE Transactions, Power Apparatus and Systems*, **86**, 1344–1356 (1967).
2. AIEE Committee Report, 'A Method for Estimating Lightning Performance of Transmission Lines', *AIEE Transactions Pt. III*, **69**, 1187–1196 (1950).
3. Harder, E. L., and Clayton, J. M., 'Transmission Line Design and Performance Based on Direct Strokes', *AIEE Transactions Pt. I*, **68**, 439–449 (1949).

4. Westinghouse Electric Corp., *Transmission and Distribution Reference Book*, 4th edn., East Pittsburgh, Pa., (1950).
5. Clayton, J. M., and Young, F. S., 'Estimating Lightning Performance of Transmission Lines', *IEEE Transactions, Power Apparatus and Systems*, **83**, 1102–1110 (1964).
6. General Electric Company, *EHV Transmission Line Reference Book*, Edison Electric Institute, New York, N.Y., (1968).
7. Fisher, F. A., Anderson, J. G., and Hagenguth, J. H., 'Determination of Lightning Response of Transmission Lines by Means of Geometrical Models', *AIEE Transactions, Pt. III*, (*Power Apparatus and Systems*, **78**, 1725–1735 (Feb. 1960 Section)).
8. Anderson, J. G., 'Monte Carlo Computer Calculation of Transmission Line Lightning Performance', *AIEE Transactions, Power Apparatus and Systems*, **80**, 414–419 (1961).
9. Anderson, J. G., and Barthold, L. O., 'Metifor, a Statistical Method of Insulation Design of EHV Lines', *IEEE Transactions, Power Apparatus and Systems*, **83**, 271–280 (1964).
10. Lusignan, J. T., and Miller, C. J., 'What Wood May Add to Primary Insulation for Withstanding Lightning', *AIEE Transactions, Power Apparatus and Systems*, **59**, 534–540 (1940).
11. Limbourn, G. J., 'Impulse Voltage Distribution between Wood and Porcelain in Transmission Line Insulation', *Institution of Engineers, Australia, Electrical Engineering Transactions*, **EE3**, 193–206 (1967).
12. Sargent, M. A., and Darveniza, M., 'The Calculation of Double-Circuit Outage Rates of Transmission Lines', *IEEE Transactions, Power Apparatus and Systems*, **86**, 665–678 (1967).
13. Sargent, M. A., and Darveniza, M., 'The Lightning Performance of Double-Circuit Transmission Lines', *IEEE Transactions, Power Apparatus and Systems*, **89**, 913–920 (1970).
14. Darveniza, M., 'Some Aspects of Lightning Flashover and Power Outage Performance of Wood-insulated Transmission Lines', *Electrical and Mechanical Transactions, Institution of Engineers, Australia*, **EM4**, No. 11, 49–60 (1962).
15. Ekvall, H. N., 'Minimum Insulation Levels for Lightning Protection of Medium Voltage Lines', *AIEE Transactions*, **60**, 128–132 (1941).
16. Gillies, D. A., 'Operating Experience with 230 kV Automatic Reclosing on Bonneville Power Administration System', *AIEE Transactions Pt. III*, **74**, 1692–1694 (1955).

6

The switching surge design of transmission lines

6.1 GENERAL

For system voltages below approximately 300 kV, switching overvoltages are not critical. The exact voltage at which switching surges
become the more important criterion depends on many factors but
mainly on the switching surge level. For this reason it is desirable to
reduce the per unit switching surges for increasing system voltages.
As a guide to a balanced design the following values have been found
satisfactory in average conditions.

Highest system voltage, kV	Maximum switching surge level, p.u.
362–420	2.5
525	2.25
765	2.0

In the design of an economical tower it is necessary to take account
of the proximity effect of grounded metal which accentuates the already
unfavourable 'saturation' characteristic of large air gaps. The effect
is illustrated in *Figure 6.1* for the centre phase of a 735 kV tower; for
outer phases it is less severe. The main problem is the co-ordination of
string length and clearances to tower legs and truss. It is usually based
on a dry positive switching impulse with a wave front giving minimum
CFO.

The next point to consider is how to relate the discharge voltage
distribution of the tower to the switching surge distribution. The
'conventional' method consists simply in fixing the composite withstand voltage of the phase conductors at some suitable margin above
the maximum switching overvoltage. The former is taken as CFO under
adverse weather conditions minus two or three standard deviations,
and the latter must be determined by analytical studies or field tests.

This procedure does not, however, take advantage of the probability that the maximum overvoltage will occur rarely, and the even lower probability of it coinciding with the assumed withstand value and unfavourable weather conditions.

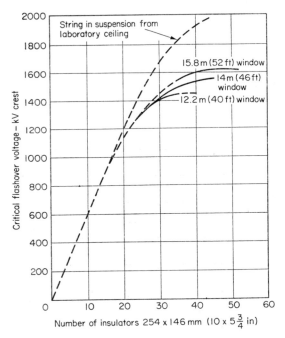

Figure 6.1. Proximity effect of tower window on insulation strength of single 90° V-strings (after Aubin et al.[5])

In the statistical method the design is based on a given risk of flashover which can be calculated by combining the flashover voltage distribution function of an insulation structure with the overvoltage probability 'density' function (see 3.2.4). The latter gives the probability $p_0(V) \, dV$ of a surge of a value between V and $V + dV$ occurring. Multiplication by the probability $P_d(V)$ of flashover at this voltage gives the probability of both events occurring simultaneously. The integral over the whole voltage range is the risk of failure, represented by the hatched area in *Figure 6.2a*. If a number of strings are subjected to the same switching surge, the distribution $P_d(V)$ has to be adjusted according to equation 3.10 or using *Figure 6.6*. The risk equals the probability of flashover per switching operation. It can also be expressed as a switching surge flashover rate, provided an estimate of the number of switching operations per year can be made.

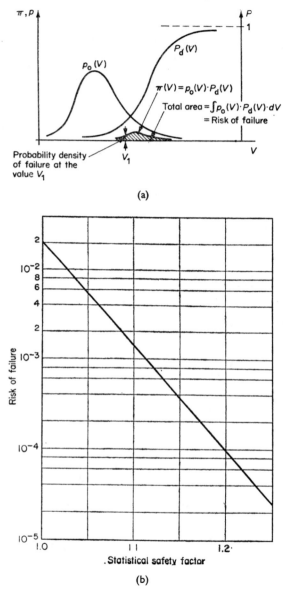

(a)

(b)

Figure 6.2. Risk of failure evaluation (after IEC Document 35, Technical Committee 28 (1970)):

(a) computation procedure;
(b) correlation between risk of failure and statistical safety factor

In the simplified method proposed by IEC[1], the risk is related to a 'statistical safety factor' which is defined as the ratio of statistical overvoltage to statistical withstand voltage. The former is the voltage likely to be exceeded by 2% of all overvoltages, the latter was already defined in Chapter 3 by a reference probability of 90%, i.e. as (CFO-1.3σ) for a normal distribution. If the risk is plotted against safety factor, as in *Figure 6.2b*, the safety factor for a desired risk can be read from this graph. The merit of this method is that it uses the conventional concept of a safety factor in a manner which permits a quantitative evaluation of the risk of a disruptive discharge as a flashover rate. The graph of *Figure 6.2b* has been conservatively computed for a discharge voltage distribution with a factor of variation of 0.08 (rather than the 0.06 suggested in Chapter 3), and for a switching surge distribution with a factor of 0.05. It should be noted that the switching surge distribution is only 'piecewise' normal between the maximum value and the minimum value of 1 p.u., and that the use of a normal distribution for the calculation of the risk gives a conservative result. The better the switching surges are controlled, the lower will be their standard deviation.

The IEC method becomes difficult to apply when many statistical distributions have to be combined. A Monte Carlo[2] computer program, based on random selection of all relevant parameters including weather effects, deals with this situation more conveniently (see 5.3.2).

Example 6.1 Estimate the switching surge flashover rate of a 525 kV line by the simplified IEC method. The maximum switching overvoltage is 2.2 p.u. with 1% probability, the p.u. standard deviation of the approximately normal distribution is 0.05 and 100 insulator strings are subjected to every surge (attenuation neglected). The composite CFO is 1280 kV for all phases; p.u. standard deviations = 0.08.

Solution: Compute the statistical safety factor as follows: maximum switching surge = 2.2 (525) $\sqrt{2/3}$ = 945 kV. Convert to 2% probability, using the normalised variable z of the cumulative normal distribution for 1 and 2% probability respectively. CFO = 945/ $(1+2.33\times0.05)$ = 845 kV; statistical overvoltage = CFO $(1+2.05 \times0.05)$ = 932 kV. Statistical withstand voltage, converted to 100 parallel insulator strings, using *Figure 6.6*, is 1280 $(1-3.07\times0.08)$ = 965 kV. Statistical safety factor = 965/932 = 1.035. The probability of outage per switching surge from *Figure 6.2b* equals 8×10^{-3}. For, say, 25 switching operations per year, the flashover rate would be 0.2 or one flashover every 5 years.

6.2 TOWER INSULATION DESIGN

6.2.1 Tower clearances

For many new and important e.h.v. projects, the tower dimensions and withstand voltages were determined by special tests on tower mock-ups and full scale towers, the tower window being the most critical. This work is fully reported in the literature. Some characteristic papers are[3,4] for 345 and 500 kV, and[5,6] for the 750 kV systems in Canada and the USA.

A preliminary slide-rule design is made possible by the graphs of *Figures 6.3–6.5* which were obtained in numerous tests at the General

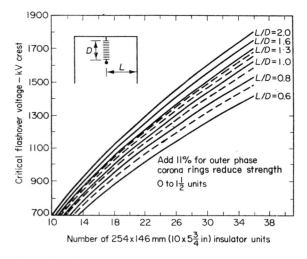

Figure 6.3. Curves for dry positive-polarity switching-surge flashover along tangent insulator strings in centre phase of steel towers (courtesy Edison Electric Institute[7])

Electric/Edison Electric Institute EHV Project[7]. Their application is shown below.

Example 6.2 Preliminary insulation design of a 345 kV, single-circuit steel tower, with free-swinging string, no contamination. Maximum switching overvoltage-to-ground (from TNA study) = 2.2 p.u. = 620 kV. The procedure is to assume a string length and check whether it is adequate.

Solution:
Step 1. Assume a string of 18 standard units, 254×146 mm (10×
$5\frac{3}{4}$ in), length $D = 2.63$ m. Allow 0.3 m for hanger and 0.3 m for
line fittings; corona rings omitted.
Step 2. Switching surge clearance to tower leg. From experience
with Metifor studies, it is recommended[7] that the clearance to tower
leg should be determined for a swing angle of 15° (which is exceeded
only 15% of time). At this angle, the clearance is assumed equal to
string length; hence dimension (L) in *Figure 6.3 is*

$$L = 2.63+(2.63+0.3)\sin 15° = 3.39 \text{ m.}$$

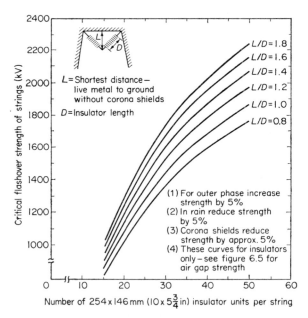

*Figure 6.4. Curves for dry positive-polarity switching-surge flashover strength of
V-string insulators (courtesy Edison Electric Institute[7])*

Step 3. Power frequency voltage should be withstood at maximum
swing angle of 60°.
    Clearance $= L-(D+0.3)\sin 60° = 0.85$ m. From *Figure 3.7,*
power frequency CFO $= 450$ kV crest at standard atmospheric
conditions. Reduce by 20% for most unfavourable air density and
humidity, and by $2\sigma = 6\%$ to obtain withstand voltage of 338 kV.
Normal line-to-ground voltage is $345\sqrt{2/3} = 272$ kV, hence there
is an ample margin.
Step 4. Find switching surge CFO voltage at 15° swing angle from

Figure 6.3 for 18 units and $L/D = 3.39/2.63 = 1.3$; CFO $= 1050$ kV. Reduction for rain, air density and humidity, say, 20% gives 840 kV. (A small correction for hanger length is here neglected.)

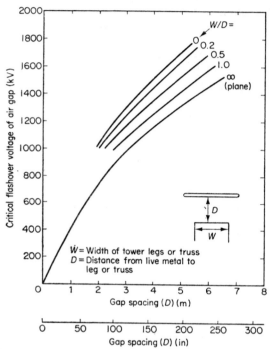

Figure 6.5. Dry positive-polarity switching-surge flashover strength of air gaps between conductor and tower leg or truss (courtesy Edison Electric Institute[7])

The withstand voltage at -2σ would be 12% lower but because a switching surge could stress a large number of towers simultaneously, a further reduction must be made. Assume that 200 towers are stressed, attenuation along the line is neglected. Enter *Figure 6.6* at the desired 97.7% withstand probability (CFO-2σ) and find at the intersection with the 200 curve that the withstand voltage is 3.7σ below the CFO of a single string, i.e. $(1-3.7\times0.06)\ 840 = 650$ kV. This is still 5% above maximum switching surge. The probability of a flashover could be estimated as in Example 6.1. A smaller number of units could perhaps be justified by more refined calculations.

Step 5. Outer strings have according to *Figure 6.3* an 11% higher CFO, hence string length and tower clearance could both be reduced for these phases.

V-strings offer a gain in right-of-way but their CFO is reduced. The greater leakage distance that can be fitted into a window of given dimensions can be an advantage in contaminated conditions. *Figure 6.4* gives guidance regarding the co-ordination of V-strings. The clear-

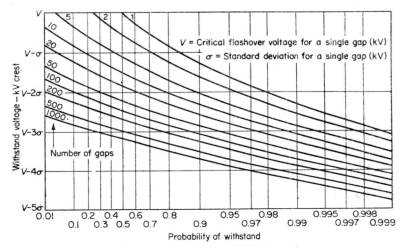

Figure 6.6. Conversion chart for reduction of withstand voltage due to placement of gaps or strings in parallel (courtesy Edison Electric Institute[7])

ances to tower legs and truss are affected by their longitudinal dimension (W), as shown in *Figure 6.5*.

The withstand voltage of the configuration can be obtained by combining the flashover distribution curves of all feasible paths in the window (along the strings and to the tower legs and truss) and then choosing from the resultant curve the voltage corresponding to the desired withstand level (*see*[7], pp. 254–255).

6.2.2 Ground clearance

For the critical flashover condition with conductor positive, the field at the ground surface is relatively weak. This explains the observation that even large objects, e.g. vehicles, protruding from the ground plane do not greatly influence the positive flashover values. CFO voltages from conductors to grounded objects underneath are given in *Figure 6.7*. For conversion to withstand and correction for variations in relative air density and humidity, a factor of 1.19 is recommended[7]. Considering that clearances to ground involve human safety, the same source suggests a further margin of 20 to 50% to allow for unknowns.

7*

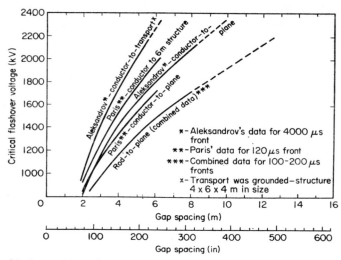

Figure 6.7. Dry positive-polarity switching-surge flashover strength between conductor and grounded object underneath (courtesy Edison Electric Institute[7])

6.3   APPLICATION TO FUTURE ULTRA-HIGH VOLTAGES

A great amount of laboratory and theoretical work is in progress for the purpose of assessing the feasibility and economy of system voltages in the range from 1000 to 2000 kV, and for acquiring the required design data[8-11]. The vital parameter, the rod-to-plane CFO voltage, is shown in *Figure 6.8* for gap distances to 30 m[11]. Conductor-to-tower CFO voltages are about 15% higher. For large gaps, the time to crest of the critical switching surge increases, as shown by the lower curve, and there is some evidence that the standard deviation also increases[9], hence the withstand voltage, for equal withstand probability, is lowered by an increasing percentage. The conclusion drawn from these curves is that the overvoltage factor must be reduced as system voltage increases.

At 765 kV, a switching surge level of 2 p.u. requires a conductor-to-tower clearance of about 5.3 m; at 1500 kV the corresponding figure is 19 m. If, however, the surge level is reduced to 1.5 p.u., which is feasible by the measures discussed in 2.3, a clearance of 11 m would be sufficient. The clearance could be further reduced to 8 m at 1.35 p.u. maximum surge. This very low surge level could probably be achieved for energising transients, but fault initiating overvoltages and contamination conditions may set the limit at about 1.5 p.u.

Ground clearance would be about 18.5 m. Allowing for vehicles and safety margins, a crossarm height of about 48 m is required. With such huge tower dimensions, every possible economy has to be explored. One way would be to ascertain, by extended analysis, the exact shape of all likely switching surges and to design for actual waveshapes instead of the most onerous. A start in this direction has been made by Clerici *et al.*[12] who have investigated gap strength for certain composite waveshapes.

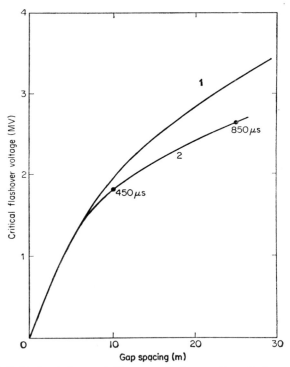

Figure 6.8. Critical rod-plane switching surge flashover voltages for large gap spacings (after Barnes and Thorén[11]):
1, time to crest 250 μs
2, times to crest for minimum CFO

REFERENCES—CHAPTER 6

1. International Electrotechnical Committee (IEC), Technical Committee, No. 28, Insulation Co-ordination, *Documents 35, 35A* (Central Office), (1970).
2. Anderson, J. G., and Thompson, R. L., 'The Statistical Computation of Line Performance Using METIFOR', *IEEE Transactions, Power Apparatus and Systems,* **85,** 677–686 (1966).

3. Alexander, G. W., and Armstrong, H. R., 'Electrical Design of a 345 kV Double Circuit Transmission Line Including the Influence of Contamination', *IEEE Transactions, Power Apparatus and Systems*, **85**, 656–665 (1966).
4. Guyker, W. C., Hileman, A. R., and Wittibschlager, J. F., 'Full-scale Tests for the Allegheny Power System 500 kV Tower Insulation System', *IEEE Transactions, Power Apparatus and Systems*, **85**, 614–623 (1966).
5. Aubin, J., McGillis, D. T., and Parent, J., 'Composite Insulation Strength of Hydro-Quebec 735 kV Towers', *IEEE Transactions, Power Apparatus and Systems*, **85**, 633–648 (1966).
6. Hauspurg, A., Caleca, V., and Schlomann, R. H., '765 kV Transmission Line Insulation: Testing Programme', *IEEE Transactions, Power Apparatus and Systems*, **88**, 1355–1365 (1969).
7. General Electric Company, *EHV Transmission Line Reference Book*, Edison Electric Institute, New York, N.Y., (1968).
8. Dillard, J. K., Clayton, J. M., and Kilar, L. A., 'Controlling Switching Surges on 1100 kV-Transmission Systems', *IEEE Transactions, Power Apparatus and Systems*, **89**, No. 8, 1752–1759 (1970).
9. Dillard, J. K., and Hileman, A. R., 'Switching Surge Performance of Transmission Systems', *Proceedings CIGRE*, Report 33-07 (1970).
10. Barnes, H. C., and Winters, D. E., 'UHV Transmission Design Requirements —Switching Surge Flashover Characteristics of Extra Long Air Gaps', *IEEE Transactions, Power Apparatus and Systems*, **90**, No. 4., 1579–1589 (1971).
11. Barnes, H. C., and Thorén, B., 'Three Years Results from the AEP-ASEA UHV Research Project', *Proceedings CIGRE*, Report 31–03 (1972).
12. Clerici, A., Colombo, A., Comellini, E., and Taschini, A., 'Considerations on the Air Insulation Design to Switching Surges in Future E.H.V. Systems', *Proceedings CIGRE*, Report 33-10 (1970).

The insulation co-ordination of high-voltage stations

7.1 PRINCIPLES

The insulation design of high-voltage stations must be based on different principles from those applying to transmission lines[1, 2]. Firstly, stations generally contain transformers and other valuable equipment with non-self-restoring insulation which must be guarded most carefully against internal breakdowns. Secondly, since they have vital functions to fulfil in the power system, even the risk of flashover in air, with the accompanying disturbance to normal operation, must be kept to a minimum.

In important stations, protection against lightning surges requires the establishment of a protective voltage level by means of shunt-connected protective devices. The lightning impulse withstand levels (BIL) of the various items of equipment have to be above the protective level by a suitable margin which, as we shall see, increases with the distance of the equipment from the protective device. This margin can be determined in respect to air insulation by the statistical methods already discussed but for non-self-restoring insulation, whose statistical withstand voltage cannot be known, the margin is chosen by the conventional method.

Below about 300 kV system voltage, switching surges are as little a critical factor in stations as on transmission lines. If the BIL is chosen correctly relative to the prevailing protective level, the equipment will also have adequate switching surge strength. In the higher voltage range, either of two policies can be followed.

(1)   Whenever possible, the sparkover voltage of the protective device should be selected so that it will not operate on switching overvoltages which, because of their relatively long duration, may cause thermal overload and damage to the device. This condition may force up the BIL.

(2)   For very high voltages, it is therefore economically desirable to use a protective device for the limitation of both switching and

lightning overvoltages. Owing to the progress in surge diverter technology this is indeed feasible. Experience with this technique is still limited and many engineers prefer to assign to the equipment a switching impulse insulation level (SIL) with a small margin above the controlled switching surge level so that surge diverters would operate on switching surges only rarely, e.g. when any of the control devices fail. They would be a second line of defence.

7.2   OVERVOLTAGE PROTECTIVE DEVICES

Only rod gaps and surge diverters (lightning arresters) will be considered, disregarding devices of more or less historical interest. Attention will be focussed on surge diverters whose characteristics can be accurately tailored to requirements.

7.2.1   Function of shunt-connected overvoltage protective devices

The withstand behaviour of insulation to surges of various types can be described by a composite voltage v. time characteristic, as

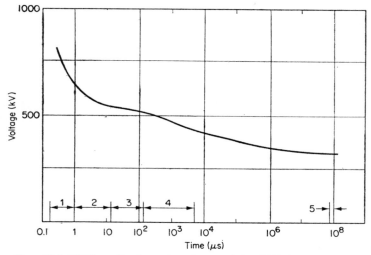

Figure 7.1. 'Withstand' voltage-time characteristic of 132 kV transformer.
Time regions:
1, steep-fronted lightning surges;
2, slow-fronted lightning surges;
3, fast switching surges;
4, slow switching surges;
5, 1 min power frequency test

shown in *Figure 7.1*, in which various time regions are indicated, each referring to a particular type or shape of test voltage. For self-restoring insulation, the withstand voltages are determined at some suitably chosen multiple of the standard deviation below the critical flashover *v.* time characteristic; for non-self-restoring insulation, points on the withstand characteristic correspond to specified test voltages for full wave, chopped wave, front-of-wave, switching impulse and power frequency.

Protection is achieved in any time region in which the protective characteristic lies below the withstand characteristic of the insulation. This is illustrated in *Figure 7.2*: the 120 kV surge diverter (Curve D)

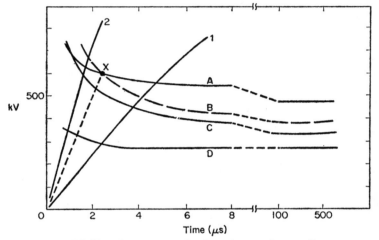

Figure 7.2. Transformer protection by rod gap and surge diverter:
  A, 550 kV BIL transformer withstand;
  B, 79 cm (26 in) rod gap flashover;
  C, 67 cm (22 in) rod gap flashover;
  D, 120 kV surge diverter sparkover;
  1,2, surge fronts

will protect the 132 kV transformer (Curve A) over the whole time range shown; the 66 cm (26 in) rod gap (Curve B) will protect the transformer only against surges with front slopes less than OX (see for example the surge whose front is shown by Line 1). Steeper surges (Line 2) will break down the insulation before the gap has had time to operate. A 56 cm (22 in) rod gap (Curve C) would improve matters but is liable to spark over for some switching surges.

The ideal requirements for shunt-connected protective devices can be summarised as follows:

(1) They must not spark over under temporary voltages under any but the most exceptional circumstances.

(2) Their volt/time curve must lie below the withstand level of the protected insulation in any time region in which protection is needed. The margin between the two curves must be adequate to allow for the effects of distance, polarity, variations in relative air density, humidity, ageing of the insulation and likely changes in the characteristics of the protective device.
(3) They must be able to discharge high-energy surges without changes in their protective level or damage to themselves or adjacent equipment.
(4) After discharging a surge they should reseal (i.e. become non-conducting) in the presence of temporary overvoltages. (Surge diverters can be destroyed by power follow current.)

7.2.2   Rod gaps

Rod gaps are simple and cheap but do not meet all requirements, in particular that of resealing. They are subject to atmospheric conditions and respond differently to surges of positive and negative polarities; their volt/time characteristic is therefore quite a broad band and bends up sharply for short wave fronts. As a consequence, their application for overvoltage protection suffers from the disabilities already illustrated in connection with *Figure 7.2*. Moreover, as no current-limiting resistor is used, the impulse voltage collapses on sparkover to zero; if the gap is installed near a transformer, the winding can be subjected to a very large step impulse which places dangerous stresses on the turn insulation.

Nevertheless, rod gaps can provide reasonable protection where the isokeraunic level is low and the front times of lightning surges are controlled by the use of overhead ground wires and the prevention of backflashovers near the station.

7.2.3   Surge diverters

(1)   *Principle of operation*   The basic elements of a surge diverter are the sparkgap, which acts as a fast switch, and the non-linear resistor. The volt/time characteristic of the multiple gap, shown as Curve D of *Figure 7.2*, is shaped by means of resistive-capacitive grading. Pre-ionisation of the gap assists in ensuring quick and consistent sparkover. The volt/current characteristic of the resistance blocks ('valve elements'), shown in *Figure 7.3a*, is such that it permits the discharge of high surge currents with an IR drop of the order of the sparkover voltage and yet limits the subsequent power frequency current to a value the gap can interrupt. The relationship follows a law

of the form

$$I_d = kV_d^a \qquad (7.1)$$

where k and a are constants depending on material and dimensions ($a = 4$–6). *Figure 7.3b* depicts typical voltage and current waveshapes during a lightning discharge operation of a surge diverter.

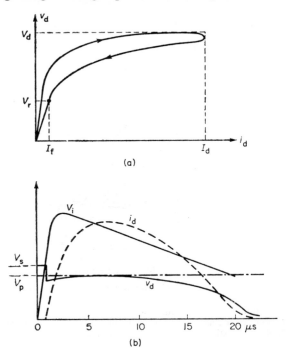

(a)

(b)

Figure 7.3. Surge diverter operation:
(a) voltage-current characteristic of non-linear resistance blocks;
(b) typical voltage and current waveshapes.
V_d = maximum voltage across diverter during discharge of a surge having a peak current of I_d;
I_f = power follow current at system voltage V_r;
V_s = sparkover voltage;
V_p = protective level

For a surge of crest voltage V, which has travelled some distance along a transmission line, the residual voltage after sparkover can be readily determined on the basis of the Thevenin-equivalent described in Appendix A3. The graphical solution is illustrated in *Figure 7.4* for two cases:
(a) line terminating at the surge diverter location, and
(b) line continuing.

In the first case, the Thevenin-voltage is $2V$ and the Thevenin-impedance is Z_0. In the second case, the corresponding values are V and $\frac{1}{2}Z_0$.

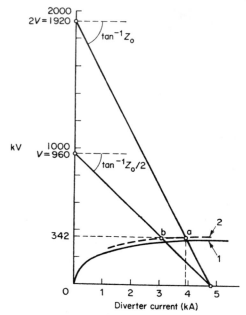

Figure 7.4. Graphical determination of discharge voltage:
(a) solution for terminating line;
(b) solution for continuing line.
1, non-linear characteristic of resistance blocks for 8/20 μs wave;
2, characteristic adjusted for steeper rate of rise of current

The equation for the diverter current

$$I_d = (V_{Th} - V_d)/Z_{Th} \qquad (7.2)$$

indicates clearly the current limiting role played by the line surge impedance. An idea of the current magnitudes involved in the worst case of a terminating line can be quickly formed from $I_d < 2V/Z_0$ by neglecting diverter resistance in equation 7.2. Assuming for a 500 kV line, $V = 2000$ kV and $Z_0 = 300$ Ω, then $I_d < 13.3$ kA. *Figure 7.4* shows that for lower system voltages (see Example 7.1), diverter currents would be smaller. Field records indicate that only 1–4% of diverter currents are greater than 10 kA and that over 70% are below 2 kA.

A very different situation confronts a surge diverter in the path of a close-in stroke to a phase conductor when the current limiting effect of Z_0 is absent. The diverter current will then be approaching stroke current magnitude with correspondingly high dynamic and thermal stresses. Another case of high thermal stress is encountered when a surge diverter sparks over on a switching surge following a capacitive interruption and has to discharge the energy stored in the open-circuited line. The consequent rectangular current surge has a moderate crest but long duration of the order of twice the line travel time. Assume for example that a 400 kV line, 300 km long, $Z_0 = 320\,\Omega$, is left with a charge corresponding to 2.5 p.u. voltage. The surge diverter discharge voltage is about 300 kV for the current involved. Applying equation 7.2, the diverter current would have a magnitude of approximately $(825 - 300\ \text{kV})/320\,\Omega = 1.65\ \text{kA}$ and a duration of 2000 μs.

The lighter designs of surge diverter cannot cope with such long duration currents, and they must not be allowed to operate for line discharges. Heavy duty surge diverters with 'assisted' or 'active' gaps are not so restricted.

(2)   *Surge diverters with active gaps*   A vital property of the gap is its ability to interrupt power follow current. For any gap design, there is a maximum power follow current, of 100–300 A, which it can safely interrupt without special measures. This current must not be exceeded at the highest power frequency voltage against which the surge diverter may have to reseal; it thereby determines the voltage/current characteristic of the non-linear resistance blocks for a given material (see equation 7.1). Fewer resistance elements could be used if a higher current could be interrupted by the gap, thus achieving a lower discharge voltage characteristic. This idea led to the development of 'active' sparkgaps which use the magnetic blow-out effect to increase the maximum gap interrupting current.

For this type of diverter the gaps are so arranged that the arc burns in the magnetic field of coils excited by the power follow current[3]. During lightning discharges, the high coil voltage induced by the steep surge front sparks auxiliary gaps by-passing the coils. The auxiliary gaps extinguish as soon as power follow current starts to flow. The magnetic field, aided by the horn shape of the main gap electrodes, extends the length of the arc and drives it into an arc cooling chamber of refractory material; the increasing arc voltage rapidly forces the current to zero as shown in *Figure 7.5*. The follow current is limited by the arc voltage drop as well as the resistance elements so that fewer resistance blocks can be used. During the surge discharge, before the arc voltage develops, the lightning protective level is accordingly lowered.

A further advantage of active gap surge diverters is that they can be used for the limitation of switching surges, with corresponding insulation economies in the extra high and ultra-high voltage range. This principle has been adopted in the planning of the first 765 kV system in the USA[4] (see Example 7.2). As the arc voltage develops already during the relatively slow rise of the switching surge, care has to be taken that excessively strong gap action does not cause the diverter voltage to rise above the switching surge protective level.

It is possible to apply these diverters in situations where the temporary overvoltage exceeds the rated diverter voltage for a limited number of cycles. The permissible duration of the temporary overvoltage is inversely proportional to its magnitude and both are limited by the thermal capacity of the diverter[3, 5].

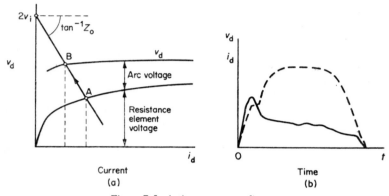

Figure 7.5. Active gap surge diverter:
(a) voltage-current characteristic;
(b) voltage and current waveshapes (current: solid line).
Point A, initial operating condition (e.g. lightning discharge);
Point B, operating condition after arc voltage fully developed

(3) *Characteristics and tests* The requirements of different systems are catered for by surge diverters of various rated discharge currents. In USA standards they are: station type (10 kA), intermediate and distribution type (5 kA) and secondary (1.5 kA). IEC recommendations provide for discharge current ratings of 10 kA (light or heavy duty), 5 kA, 2.5 kA and 1.5 kA. A brief review of the usual tests, based on USA[6] and IEC[7, 8] standards will help to convey an appreciation of the diverse requirements a surge diverter has to meet*.

(a) The power frequency sparkover test establishes the minimum margin to sparkover above rated voltage.

(b) The lightning impulse sparkover/time characteristic is obtained by testing with standard 1.2/50 µs impulses of increasing magnitude (see *Figure 7.2*). The front slope is correspondingly increased until it reaches the specification value of approximately 100 kV/µs for each 12 kV of diverter rating, with an upper limit of 1200 kV/µs. At the

* British Standard BS 2914:1954, in the process of revision (see Document No. 25918), is expected to conform substantially to the IEC recommendations.

specified front slope the front-of-wave sparkover voltage is measured.
(c) The test for switching impulse sparkover voltage/time characteristic was introduced by IEC in 1970. It applies to 10 kA surge diverters above 100 kV rated voltage. In view of the importance of switching
surges in e.h.v. installations, national standards will probably follow.
IEC recommends tests with three waveshapes of both polarities,
with front times 30–60 μs, 150–300 μs and 1000–2000 μs. The times
to half value should be appreciably longer than twice the front time
but the exact value is not considered critical.
(d) The discharge voltage characteristic establishes the maximum
discharge voltage across the surge diverter, usually for at least three
currents: 0.5, 1.0 and 2.0 times rated discharge current of 8/20 μs
shape.
(e) The high current, short duration test (100 kA for station type, 65 kA
for intermediate and distribution type, 10 kA for secondary type;
all of 4/10 μs to 8/20 μs duration) checks the electrical and mechanical strength of the diverter for nearby strokes, the ability of the blocks
to resist puncture and contour flashover, the performance of the
sparkgap setting, and the quality of the grading circuit.
(f) The low-current, long duration test proves the ability of resistor
blocks to withstand switching surge discharges or 'hot lightning'.
For distribution and light-duty type diverters, 20 discharges of 75–
250 A in a rectangular wave of 1000–2000 μs are normally used.
For heavy-duty type diverters, a more stringent test is prescribed
which simulates the discharge of a long transmission line by the
discharge of a suitably adjusted impulse current generator[9].
(g) The duty cycle test simulates operation under lightning discharge
and consists of 20 operations at rated discharge current (e.g. 10 kA,
8/20 μs for station type) with power follow current. It checks the
discharge voltage and the ability to reseal after lightning discharges.
(h) Pressure relief devices, intended to prevent explosion of the porcelain housing as a consequence of internal flashover or failure to reseal,
are often provided and are suitably tested.
(i) Proposals for pollution tests are contained in IEC Publication
99-1 and are being considered in the British Standard under revision.
Pollution can cause non-uniform distribution of the operating voltage
across the outer porcelain housing. The current surges caused by
breakdowns of 'dry bands' induce voltages in the gap assembly which
can cause sparkover in normal operation and destruction of the surge
diverter by sustained power frequency current[10]. Resistance grading
of the sparkgap is an effective remedy.

Constancy of the sparkover characteristics is vital for the correct
operation of the surge diverter. It must be assured by reliable sealing
of the housing, particularly the chambers containing the gaps. Moisture
is the enemy as it forms a corrosive acid with the usual nitrogen filling.

7.3 STATIONS WITH PROTECTED ZONE

7.3.1 Magnitude and shape of incoming voltage surges

Direct strokes to phase conductors in proximity of surge diverters would cause very high currents to flow in the diverter and the discharge voltage would be too high to permit an economical insulation design. To take advantage of the current-limiting effect of the line surge impedance, it is essential to shield important stations against direct strokes by masts or shielding wires (using the methods described in 4.3) and to extend the shielding to the incoming lines for a distance of at least 800 m, preferably 1.5–2 km (if they are not already so equipped for their whole length). With this 'protected zone', most surges arriving at the substation will have originated outside that zone.

At the tower where flashover occurs, the voltage on the conductor is determined by the stroke current and the combined impedance of ground wires, conductor and tower footing resistance. This voltage (V_T), travelling on the conductor towards the substation, is quickly attenuated to $(1 - K)V_T$, due to coupling with the shielding wire/earth loop, and in addition suffers further attenuation and distortion depending on the length of the protected zone. The voltage wave at the station entrance can be estimated by assuming a voltage magnitude at the beginning of the protected zone equal to 1.2 times the negative CFO voltage of the line insulation (the negative value being the higher one), with a vertical front as for a backflash, and applying to it the rules of attenuation and distortion given in 4.7. It will be seen below that the sloping off of the surge front is important for the effectiveness of surge diverters.

The possibility of a backflash inside the protected zone cannot be excluded. For such a rare contingency, the risk of a more severe surge and some encroachment on the safety margin must be accepted. The probability of this occurring depends on the backflashover rate of the protected zone and its length. For a station with n lines, this probability can be kept constant by reducing the protected zone of each line to one nth. The maximum steepness of the incident wave increases, because of the shorter travel, but at the busbar steepness and magnitude are reduced by the lower surge impedance of $(n-1)$ parallel lines (see 7.3.9).

7.3.2 Rated diverter voltage and protective level

The rated diverter voltage is normally so chosen that it is not less than the temporary overvoltage to ground at the point of installation under any reasonably likely fault or abnormal operating condition (see 2.2). For any particular type and make of surge diverter with the selected rated voltage, the sparkover and discharge voltages can be ascertained from catalogues or tenders, and the protective level at the surge diverter terminals derived (see *Table 7.1*). For simplified calculations, the protective level is taken as the highest of the following voltages: standard wave impulse flashover, discharge voltage and front-of-wave sparkover divided by 1.15. This reduction compensates approximately for the short duration of the voltage spike before sparkover.

Table 7.1 TYPICAL CHARACTERISTICS OF SURGE DIVERTERS IN RANGE 100–200 kV, 10 kA, LIGHT AND HEAVY DUTY

	Per unit (*referred to rated diverter voltage* r.m.s.)
Maximum 1.2/50 µs sparkover	2.2–2.8
Maximum front-of-wave sparkover (at specified rate of rise)	2.9–3.1
Maximum switching impulse sparkover	2.3–3.0
Maximum discharge voltage for 8/20 µs current wave	
1.5 kA	1.6–2.2
5 kA	2.0–2.7
10 kA	2.2–3.0
20 kA	2.5–3.3

7.3.3 Equipment insulation level

For steep-fronted travelling waves the voltages at different points in the substation exceed the protective level by amounts depending on the distance from the diverter location, the steepness of the wave front and the electrical parameters of the station. The next task is therefore to decide on the number and locations of diverters which optimise overall cost. In substations of moderate physical dimensions, a judicious choice of location can keep the number of diverters to a minimum. In the higher voltage range, it is usual to install diverters between a transformer and its circuit-breaker in order to protect the former from overvoltages caused by current chopping. Furthermore, a location nearest the transformer offers the best prospect of economies, because of the high cost of transformer insulation.

The basic lightning insulation level (BIL) is often determined by simply adding a margin, say 25–30%, to the protective level of the surge diverter and then selecting the next higher BIL from the list of standard values (*Tables 3.3* and *3.4*). This rough and ready approach can be justified for stations of minor importance; in large and important stations it is necessary to allow for the 'distance effect' more accurately. Hand calculations are practicable only in simple cases; conservative answers can be obtained by various estimating methods and pre-calculated graphs[11, 13, 14]. Extra high voltage installations call for special TNA or computer studies if the optimum solution in the choice of insulation levels is sought. Depending on the degree of accuracy achieved, a smaller or larger safety margin should be added, the minimum being 10% for the possibility that the insulation quality of the equipment and/or the surge diverter characteristic may deteriorate with time.

When the surge diverter is used to establish a switching surge protective level, the margin is usually only 15%. The distance effect is negligible because of the low front steepness. A front of only 50 µs is 15 km long.

7.3.4 Distance effect

As a first approach it is instructive to study the voltages at various points of a simple in-line arrangement of circuit-breaker, surge diverter and transformer, which may represent an emergency condition of a more complicated substation, for an incoming flat-topped surge of constant front steepness S. The transformer is approximated by an open circuit, all lumped capacitances are neglected, and the surge diverter is treated, as explained in Appendix A3, as the source of a voltage-cancelling wave. Its characteristic is idealised by the chain-dotted line in *Figure 7.3b*, corresponding to a constant protective level (V_p). The voltage spike caused by a possible sparkover above V_p can be neglected.

The voltage waves for three locations are plotted in *Figure 7.6* —directly along the time axes of the lattice diagram, for easy correlation. It will be seen that the maximum voltage at an object at a distance D from the surge diverter, either upline or downline, is

$$V(D) = V_p + 2ST \qquad (7.3)$$

where $T = D/u$, the travel time between the surge diverter and the object. Irrespective of the crest of the incoming voltage, $V(D)$ cannot exceed $2V_p$; this maximum value is attained for $2T > T_0$, where T_0 = time to surge diverter sparkover. For the case that the line extends sufficiently beyond the surge diverter so that reflections from its end do not return before the overvoltages at the substation have subsided to innocuous values, the voltage downline of the diverter is limited to V_p.

A real substation is very different from the simple model discussed above. The first correction must take account of the effective surge capacitance of the transformer

$$C_{\text{eff}} = (C_g C_s)^{1/2} \qquad (7.4)$$

where C_g = winding plus bushing capacitance to ground and C_s = winding series capacitance. The transformer inductance can be neglected with small error because of its relatively large time constant (L/Z_0). The value of C_{eff} is in the range from 500 to 5000 pF; it

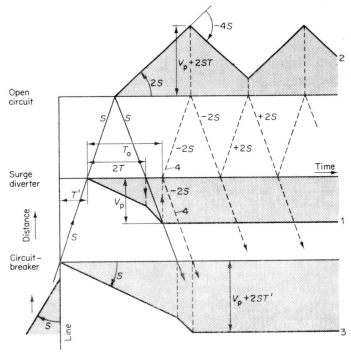

Figure 7.6. Determination of surge voltages in single-line, single-transformer station by means of lattice diagram:
1, surge diverter voltage;
2, voltage at transformer represented as open circuit;
3, voltage at circuit-breaker;
4, voltage-cancelling wave sent out by surge diverter

increases with rating and falls with BIL. Estimating values can be found[11]. The time constant $(C_{\text{eff}}Z_0)$ takes values from 0.2 to 2 µs. For the same layout as already discussed, *Figure 7.7* gives the ratio of transformer voltage V_T to protective level as a function of T/T_0. V_T increases with $(Z_0 C_{\text{eff}})/T_0$ and may even exceed $2V_p$. A method

8*

of allowing for transformer capacitance by using pre-calculated graphs was published by an IEEE Committee[11].

Lumped capacitance near the station entrance has a beneficial effect as it reduces the slope of the incident wave front. Capacitive potential transformers with their high capacitances, which range from 2000 to 8000 pF, are very effective in this respect. The result of a computer study of a 330 kV substation in its first stage, or later contingency condition, is shown in *Figure 7.8* which illustrates the voltages at different points and the effect of capacitor voltage transformers.

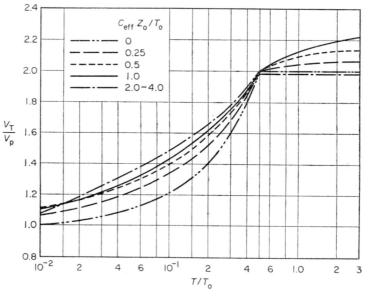

Figure 7.7. Curves for estimating transformer voltage (V_T) in single-line, single-transformer station, allowing for effective transformer capacitance (c_{eff}) (after Witzke, Bliss[18]):

V_p, surge diverter protective level;
T, travel time from surge diverter to transformer;
T_o, time to flashover of surge diverter

In computer and TNA studies it is easy to allow quite accurately for physical details of substation layout and small lumped capacitances and inductances. Lumped parameters are treated as stub lines, as explained in Appendix A2. Suspension insulators have an average capacitance of 80 pF/unit; this figure has to be divided by the number of units in the string. Similarly for pincaps which have 100 pF/unit. Busbar surge impedance can be calculated like that of an overhead line but with conductor capacitances increased by the capacitances of the insulators.

Figure 4.7 is a sample of oscillograms recorded in staged tests using a realistic steep fronted, short tailed surge of 0.6/6 μs applied at 2.2 km from the station entrance. In these tests it was established that the voltage drop in the surge diverter lead can be calculated as a *Ldi/dt* voltage with *L* = 1.2 μH/m. When added to the surge diverter discharge voltage, an 'effective protective level' is obtained, a welcome simplification for hand calculations. The rate of change of current

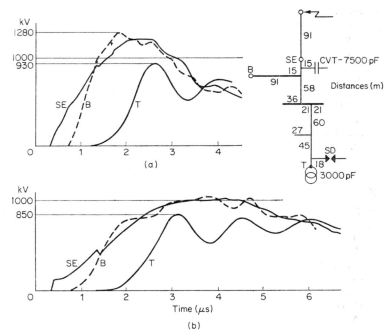

(a)

(b)

Figure 7.8. Computer calculation of surge voltages in 330 kV substation for incoming 1.2/50 μs surge, 1500 kV peak; surge diverter protective level 750 kV:

(a) without capacitor voltage transformer;
(b) with capacitor voltage transformer, 7500 pF. (Courtesy Electricity Commission of New South Wales, private communication.)

T, voltage at transformer; B, at bus end;
SE, at station entrance

can be conservatively estimated from the maximum current and the front time of the voltage wave. The discharge voltage of the surge diverter increases with increasing rate-of-rise of current. A correction of the published figures for 8/20 μs waves is advisable; it can be taken from Fig. 27 on page 625 of [20], or from the approximate correction factor

$$1+0.015(di/dt)/(di/dt \text{ of test wave}) \qquad (7.5)$$

The surge diverter grounding resistance does not enter the calculation provided the protected equipment is connected to the same grounding grid.

Breuer et al.[13] published complete TNA solutions for typical 138, 230 and 345 kV substations which can be applied to similar layouts.

Example 7.1   Determine the BIL of a transformer and switchgear in a 138 kV substation to which two single-circuit lines (leading in different directions) are connected.

Line insulation: 10 discs 254×146 mm ($10 \times 5\frac{3}{4}$ in)

Conductor diameter: 2.35 cm, average height 11.75 m, on steel towers

Protected zone = 1.6 km

Surge impedance = 400 Ω

Transformer effective capacitance = 1000 pF

Maximum operating voltage = 140 kV

Earthing factor = 1.35

Maximum temporary overvoltage = 1.05 p.u.

Distance surge diverter to transformer = 15 m

Surge diverter lead length = 6 m

Distance surge diverter to line entrance 60 m.

Solution:

Step 1. *Maximum surge* at beginning of protected zone = 1.2 (negative CFO of 10 discs, from *Figure 3.7*) = 1130 kV

Corona inception voltage (from equations 4.19 and 4.21) = 257 kV

Attenuated crest (from equation 4.26) = 960 kV

Front prolongation (from *Figure 4.9*) = 2 μs. Assume initial front vertical; hence slope of incident wave S = 480 kV/μs

Time to half value, assume average figure of 20 μs

Maximum temporary power frequency voltage to ground = $140(1/\sqrt{3})$ 1.35 (1.05) = 115 kV r.m.s.

Step 2. Choose lowest standard surge diverter rating above 115 kV = 120 kV

Catalogue values for a 10 kA, light duty surge diverter (highest values from *Table 7.1* assumed here):

front-of-wave sparkover, maximum = 3.1 (120) = 372 kV
full wave sparkover, maximum   = 2.8 (120) = 336 kV
switching surge sparkover maximum = 3.0 (120) = 360 kV
maximum discharge voltages for 8/20 μs current wave

kA:	1.5	3.0	5.0	10.0
kV:	264	308	336	360

Step 3. Consider a contingency situation with only one line and transformer in service. The discharge current will not exceed

2×960 kV/400 Ω = 4.8 kA. At 2 μs front time, the rate of rise of current is about four times the standard rate of rise; according to equation 7.5, add 6% to the catalogue discharge voltages. The actual discharge current and voltage are found from the graphical construction of *Figure 7.4* as

$$V_d = 342 \text{ kV and } I_d = 3.9 \text{ kA.}$$

Voltage drop in surge diverter lead = 6 (1.2 μH/m) 3.9 kA/2 μs = 14 kV. Effective discharge voltage = $342+14 = 356$ kV. Protective level is highest of: $372/1.15 = 324$ kV, 336 kV and 356 kV; hence $V_p = 356$ kV.

Step 4. *Distance effect for transformer:*

$$2ST = 2 \text{ (480 kV/}\mu\text{s) (15/300)} = 48 \text{ kV}$$

Approximate transformer voltage peak = $356+48 = 404$ kV. To provide a margin of not less than 25%, the BIL should be 550 kV. A more accurate determination, allowing for transformer capacitance, uses *Figure 7.7*:

$$C_{eff}Z_0 = 10^{-3} \text{ } \mu\text{F} \times 400 \text{ } \Omega = 0.4 \text{ } \mu\text{s}$$
$$T_0 = 356/480 = 0.7 \text{ } \mu\text{s}$$
$$T = 15/300 = 0.05 \text{ } \mu\text{s}$$

Enter *Figure 7.7* with $C_{eff}Z_0/T_0 = 0.54$ and $T/T_0 = 0.0675$ and find $V_T/V_p = 1.34$, $V_T = 1.34$ (356) = 477 kV. The margin is now only 15% for a BIL of 550 kV but should be acceptable in view of the more accurate calculation.

Step 5. *Transformer switching impulse withstand =*
$$0.83 \text{ (BIL)} = 456 \text{ kV.}$$
Required margin above switching surge sparkover of 360 kV is 15%; actual margin 27%.
Minimum switching surge sparkover is about 0.9 of maximum = 322 kV, corresponding to 2.8 p.u. switching surge; this should ensure rare operation under switching surges.

Step 6. *Distance effect for line entrance* = 2ST = 192 kV.
Voltage peak at entrance = $356+192 = 548$ kV.
An appropriate standard BIL would be 650 kV.

Step 7. *Circuit-breaker switching surge withstand* can be estimated at 0.65 (BIL) = 423 kV. Actual margin = 17%.

Example 7.2 The 765 kV stations of the American Electric Power Service Corporation are protected by surge diverters against both lightning and switching surges[4]. The diverters are capable of discharging temporary overvoltages and will withstand 10 such operations in successive half cycles of power frequency. Within five cycles, relays will trip the line thus removing the overvoltage. The diverters are liable to deteriorate if so used but will not fail immediately. Devices for monitoring their condition have been provided. If the surge diverters had been chosen in the conventional way, with a rated voltage above the maximum temporary overvoltage, a much higher BIL would have been necessary. Some of the basic data are:

Rated diverter voltage = 588 kV
Seal-off against 60 Hz plus harmonics = 950 kV crest (minimum)
Lightning surge protective level = 1440 kV
Transformer and reactor BIL = 1800 kV
Circuit-breaker and current transformer BIL = 2300 kV
Switching surge protective level = 1200 kV
Transformer and reactor SIL = 1500 kV
Circuit-breaker and current transformer SIL = 1350 kV closed, 1500 kV open
Maximum switching surge from TNA studies = 2.0 p.u. = 1250 kV.

7.3.5   Evaluation of non-standard waves

The waveshapes at various points in a station are irregular, usually consisting of unidirectional and oscillatory components. The withstand voltage of insulation in respect of such waves is not precisely known. None of the methods that have been proposed to obtain equivalence between a non-standard wave and the standard 1.2/50 μs test wave have a secure theoretical basis; for this reason the simplest, suggested by Breuer et al.[13], will be presented. It relates a non-standard wave to the 'withstand' volt/time curve of a transformer, which is really a composite of three withstand test points for (a) the standard full wave, (b) the chopped wave, and (c) the front-of-wave (*Figure 7.9*). The equivalence is considered established if the non-standard wave meets two requirements:

(1) It must remain at all points below the transformer withstand curve.

Since this condition would still admit waveshapes that stress the insulation more than the conventional proof tests, it is necessary to impose a second condition, which has been arbitrarily chosen, as follows:

(2) The surge must not be above the line AC in *Figure 7.9* for longer than half the time AC.

A number of graphs at various percentage ordinates of the basic withstand curve are drawn. To the non-standard wave is ascribed a 'percent surge factor' corresponding to the percent curve for which it meets the two conditions. A margin of 10% is required, i.e. a 90% surge factor wave is acceptable. For porcelain/air insulation, which has a steeper turn-up for short times, the transformer withstand curve can serve as a conservative substitute. The method is convenient as it permits the use of a template while other methods require complex integrations.

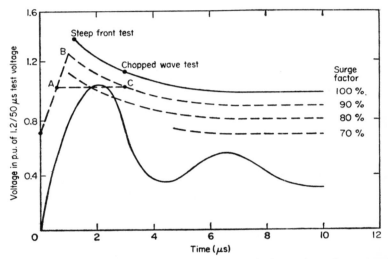

Figure 7.9. Graphical method for evaluating non-standard waveshapes by means of surge factor (after Breuer et al.[13]). Sample wave has 85% surge factor.

7.3.6 Effect of power frequency voltage

If the surge diverter sparks over at the instant when the p.f. voltage is at the maximum of opposite polarity to the surge, the surge magnitude and hence the time to sparkover are increased. This allows the surge voltage at the transformer to rise to a higher peak value. The voltage to ground stressing the transformer is this surge voltage reduced by the p.f. voltage. The net effect is an increased stress of the transformer insulation.

7.3.7 Electrical clearances in stations

In stations below 230 kV, air clearances are based on the maximum lightning surge voltage at any point, corrected for power frequency voltage. The flashover data, of which a survey is given in Chapter 3, are used with a safety margin of at least 15% to allow for variable atmospheric conditions. Many standards[2] provide schedules of recommended clearances related to equipment BIL.

For voltages from 230 kV upwards, switching surges become important. The maximum design switching surge has to be ascertained, and the SIL determined allowing three standard deviations for withstand values. For phase-to-ground clearances, rod-to-plane test data (or conductors-to-plane, if available) are used, for phase-to-phase clearances, rod-to-rod (conductor–conductor) data offer a better fit. An IEEE Committee Report[14] presents figures based on a range of switching surges from 2.0 to 2.8 p.u. for phase-to-ground and from 3.0 to 4.5 p.u. for phase-to-phase, and withstand probabilities from 90% up.

7.3.8 Station entrance protection

An open line breaker or disconnect switch presents the problem of guarding against flashover of the open contacts. Preferential flashover to ground on the line side is desirable but to achieve it by differential insulation is expensive. The provision of surge diverters on the line side, though ideal, is too costly a solution for the relatively rare occasions when the line is disconnected.

Rod gaps can be a satisfactory compromise if carefully adjusted so that there is a small probability of sparkover[15]. When the line is connected to the station they must not sparkover or they would vitiate surge diverter operation. With the line disconnected, a line outage would not interrupt power but may be an operational inconvenience. If a line is out of service for any length of time, the problem can be solved by grounding the line.

7.3.9 Self-protecting stations

The familiar Thevenin-theorem tells us that in a station with n lines connected to the busbars, an incoming surge V on one line will be reduced to $2V/n$. This assumes that the lines are sufficiently long for the reflected waves from their far ends to have no effect on the initial peak. The effect of reflections from the point of origin of the surge is also neglected. Assuming that the origin is at the beginning of the protected zone, and that the surge voltage peak impressed at the point of strike is of longer duration than the return time from the substation, the peak voltage at the busbar will take the values of *Table 7.2*[16].

The front slope is reduced in the same proportion as the peak voltage. The effect on the choice of the length of the protected zone is discussed in 7.3.1.

Advantage can be taken of the lowering of the surge magnitudes, provided a minimum number of lines can be relied upon to be in service all the time. In such cases it may be possible to omit surge diverters altogether.

Table 7.2 SURGE VOLTAGE ON BUSBARS IN P.U. OF INCOMING SURGE FOR 0/20 μs
WAVE (ATTENUATION NEGLECTED)

Number of lines	Distance to point of strike, km		
n	0.8	1.6	2.4
2	1.0	1.0	1.0
3	0.74	0.69	0.67
4	0.64	0.59	0.53

7.4 STATIONS WITHOUT PROTECTED ZONE

The absence of a shielded and well earthed protected zone, as one might find in distribution systems, implies that for close-in strokes the current-limiting effect of line surge impedance is missing and the steepness of the incoming voltage wave may be insufficiently attenuated. It is therefore essential that a surge diverter should be located at the transformer terminals and the lead length eliminated in order to suppress the distance effect. The choice of BIL should be based on the higher discharge voltages expected for the higher diverter currents and front steepness. There is, however, an economic limit to the expenditure on surge protection that can be justified. An efficient transformer exchange and repair service may help to keep overall costs down.

7.5 CABLE-CONNECTED EQUIPMENT

The surge impedance of a cable is of the order of one-tenth of that of an overhead line, hence a surge arriving at a line/cable junction will on first incidence be reduced to

$$(2Z_c/Z_0)/(1+Z_c/Z_0) = 0.18 \quad \text{p.u. approximately} \qquad (7.6)$$

However, the high impedance of the transformer usually connected at the other end of the cable reflects this surge practically unchanged (the transformer capacity has little influence because of the small time constant $Z_c C_{eff}$). On return to the line/cable junction, the surge is reflected in the ratio

$$(1-Z_c/Z_0)/(1+Z_c/Z_0) = 0.82 \quad \text{p.u. approximately} \qquad (7.7)$$

The voltage on the cable builds up as surges travel back and forth. If the surge is long, the voltage at the transformer ultimately assumes the value $2V$ as if the cable were non-existent. There is, however, an improvement in the slower increase in voltage. If the stroke is of short duration, a cable of moderate length can produce a substantial reduction in transformer surge voltage. *Figure 7.10* illustrates these phenomena.

A surge diverter at the junction operates when the voltage, after some reflections, has built up to sparkover value. The distance effect of the cable depends on cable length and surge front steepness. A surge diverter at the transformer end is more effective but is not feasible if the cable enters the tank directly. Calculation time can be saved by consulting published studies[17-19].

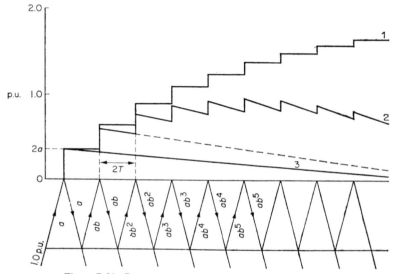

Figure 7.10. Surge voltages at cable-connected equipment:
1, for incoming step voltage wave;
2, for incoming triangular wave;
3, triangular waveshape;
a (transmission factor) = 0.18;
b (reflection factor) = 0.82

REFERENCES—CHAPTER 7

1. International Electrotechnical Commission, (IEC), *Insulation Co-ordination*, Publication 71, 4th edn. (1967).
2. International Electrotechnical Commission, (IEC), *Application Guide (Supplement to Publication 71)*, Publication 71A (1962). See also drafts for new edition, Documents 28 (Secretariat), 65 and 66 (May, 1972).

3. Sakshaug, E. C., 'Current Limiting Gap Arresters—Some Fundamental Considerations', *IEEE Transactions, Power Apparatus and Systems*, **90**, No. 4, 1563–1573 (1971).
4. Phelps, J. D. M., Pugh, P. S., and Beehler, J. E., '765 kV Station Insulation Coordination', *IEEE Transactions, Power Apparatus and Systems*, **88**, 1377–1382 (1969).
5. Flugum, R. W., 'Operation of Lightning Arresters on Abnormal Power Frequency Voltages', *IEEE Transactions, Power Apparatus and Systems*, **89**, No. 7, 1444–1451 (1970).
6. American Standard ANSI/IEEE C62.1—1967, *Lightning Arresters for A.C. Power Circuits* (being revised).
7. International Electrotechnical Commission (IEC), *Lightning Arresters*, Publication 99-1 (1970).
8. International Electrotechnical Commission (IEC), *Application Guide (Supplement to Publication* 99-1, 1958), Publication 99-1A (1965) (Amended in Appendix C of Reference 7).
9. Yost, A. G., Carpenter, T. J., Links, G. F., Stoelting, H. O., and Flugum, R.W., 'Transmission Line Discharge Testing for Station and Intermediate Lightning Arresters', *IEEE Transactions, Power Apparatus and Systems*, **84**, 79–87 (1965).
10. Torseke, L., and Thorsteinsen, T. E., 'The Influence of Pollution on the Characteristics of Lightning Arresters', *Proceedings CIGRE*, Report 404 (1966).
11. IEEE Committee Report, 'Simplified Method for Determining Permissible Separation between Arresters and Transformers', *IEEE Transactions, Power Apparatus and Systems*, **82S**, 35–55 (1963).
12. Gross, I. W., Griscom, S. B., Clayton, J. M., and Price, W. S., 'High-Voltage Impulse Tests in Substations', *IEEE Transactions, Power Apparatus and Systems*, **73**, 210–220 (1954).
13. Breuer, G. D., Hopkinson, R. H., Johnson, I. B., and Shultz, A. J., 'Arrester Protection of High-Voltage Stations against Lightning', *IEEE Transactions, Power Apparatus and Systems*, **79**, 414–423 (1960).
14. IEEE Committee Report, 'Minimum Electrical Clearances for Substations Based on Switching Surge Requirements', *IEEE Transactions, Power Apparatus and Systems*, **84**, No. 5, 415–417 (1965).
15. Hileman, A. R., Wagner, C. L., and Kisner, R. B., 'Open-breaker Protection of EHV Systems', *IEEE Transactions, Power Apparatus and Systems*, **88**, 1005–1014 (1969).
16. Clayton, J. M., and Young, F. S., 'Application of Arresters for Multi-line Substations', *IEEE Transactions, Power Apparatus and Systems*, **79**, 566–575 (1960).
17. IEEE Committee Report, 'Surge Protection of Cable-connected Distribution Equipment on Underground Systems', *IEEE Transactions, Power Apparatus and Systems*, **89**, 263–267 (1970).
18. Witzke, R. L., and Bliss, T. J., 'Surge Protection of Cable-connected Equipment', *AIEE Transactions, Power Apparatus and Systems*, **75**, 1381–1386 (1957).
19. Powell, R. W., 'Lightning Protection of Underground Residential Distribution Circuits', *IEEE Transactions, Power Apparatus and Systems*, **86**, 1052–1056 (1967).
20. Westinghouse Electric Corp., *Transmission and Distribution Reference Book*, 4th edn., East Pittsburgh, Pa., (1950).

Bibliography

Bewley, L. V., *Travelling Waves on Transmission Systems*, Dover Publications, N.Y., (1963).

Diesendorf, W., *Overvoltages on High Voltage Systems* (*Insulation Design of Transmission Lines and Substations*), Rensselaer Bookstore, Troy, N. Y., (1971).

Forrest, J. S., Howard, P. R., Littler, D. J., (ed.), *Gas Discharges and the Electricity Supply Industry*, Proceedings of the International Conference, Leatherhead, England, May 1962, Butterworths, London (1962).

Greenwood, A., *Electrical Transients in Power Systems*, Wiley, New York, (1971).

Lewis, W.W., *Protection of Transmission Systems against Lightning*, Dover Publications, New York, (1965).

Malan, D. J., *Physics of Lightning*, The English Universities Press, (1963).

Peterson, H. A., *Transients in Power Systems* Dover Publications, N.Y., (1963).

Rudenberg, R., *Transient Performance of Electric Power Systems*, McGraw-Hill, N.Y., (1950),

Rudenberg, R., *Electrical Shock Waves in Power Systems*, Harvard University Press, Cambridge, Mass., (1968).

Schonland, B. F. J., *Atmospheric Electricity*, 2nd end., Methuen, (1953).

Willheim, R., Waters, M., *Neutral Grounding in High-voltage Transmission*, Elsevier, London, (1956).

Appendix A

Propagation of travelling waves

A1. TRAVELLING WAVES ON SINGLE CONDUCTOR/GROUND LOOP

A1.1   A charge $q_0 \Delta x$ placed on a line element Δx by any process whatever is accompanied by a voltage $v = q_0/C$ where q_0 is the charge and C is the capacitance per unit length. If the charge is at the beginning of a long line, it will propagate as a travelling wave along the line at a velocity

$$u = (LC)^{-1/2} \qquad \text{(A.1)}$$

where L is the inductance per unit length. On overhead lines, u is slightly below the velocity of light; for approximate calculations a value of 300 m/μs (1000 ft/μs) can be used. On insulated cables the velocity has approximately half this value.

A1.2   At any point at which the charge passes, the current is given by

$$i = q_0 u = vC(LC)^{-1/2} = v(L/C)^{-1/2} = v/Z_0 \qquad \text{(A.2)}$$

Z_0 is the surge impedance of the line.

On a loss-free line, voltage and current waves propagate without change of shape or magnitude.

A1.3   If the charge originates in the middle of a line, half of it propagates in each direction, hence:

$$v = v' = q_0/2C \qquad \text{(A.3a)}$$

$$i = -i' = v/Z_0 \qquad \text{(A.3b)}$$

The prime indicates the wave in the negative direction. Note that the two voltage waves have the same polarity whilst the two current waves have opposite signs; this is so because i' flows against the assumed positive direction.

A1.4   A travelling wave (v, i) which encounters an impedance discontinuity, generates a reflected wave (v', i') travelling in the opposite

119

direction. If the discontinuity is an impedance Z_2 terminating a line of surge impedance Z_1, the boundary conditions at the terminating node are

$$\text{node voltage} \qquad V_n = v + v' \qquad\qquad \text{(A.4a)}$$

$$\text{node current} \qquad I_n = i + i' \qquad\qquad \text{(A.4b)}$$

$$V_n/I_n = Z_2 \qquad\qquad \text{(A.4c)}$$

and may be represented by the Thevenin-equivalents shown in *Figure A.1.*

$$V_n = v2Z_2/(Z_2+Z_1) \qquad\qquad \text{(A.5a)}$$

$$I_n = i2Z_1/(Z_2+Z_1) \qquad\qquad \text{(A.5b)}$$

$$v' = v(Z_2-Z_1)/(Z_2+Z_1) = \Gamma v \qquad\qquad \text{(A.6a)}$$

$$i' = i(Z_1-Z_2)/(Z_1+Z_2) = -\Gamma i \qquad\qquad \text{(A.6b)}$$

in which Γ is the 'reflection coefficient'.

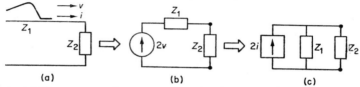

Figure A.1. Thevenin equivalent circuits (b) and (c) for surge (v, i) travelling on a transmission line (a) of surge impedance (Z_1) and impinging on a termination of impedance, or surge impedance, (Z_2)

If the impedance discontinuity is a junction with a line of different surge impedance Z_2, the same relations apply, and the voltage and current developed at the node propagate as a transmitted wave (v'', i'') into the second line, so that

$$v'' = v2Z_2/(Z_2+Z_1) = v(1+\Gamma) \qquad\qquad \text{(A.7a)}$$

$$i'' = i2Z_1/(Z_2+Z_1) = i(1-\Gamma) \qquad\qquad \text{(A.7b)}$$

The factors in equations A.7 are the voltage and current 'transmission coefficients'.

It will be seen that

$$v/i = Z_1, \quad v'/i' = -Z_1 \quad \text{and} \quad v''/i'' = Z_2$$

A2. REPRESENTATION OF LUMPED CAPACITANCE AND INDUCTANCE

A2.1   The shapes of reflected and transmitted waves are the same as the shape of the incident wave if the discontinuities consist of loss-free lines or resistive elements. This is no longer true if C and/or L elements are involved. Calculations are often simplified by substituting equivalent transmission line stubs for lumped elements.

A2.2   A shunt-connected lumped capacitance C_2 can be replaced by an open-circuited line of length l_s, surge impedance Z_s, unit capacitance C_s and travel time T_s. The condition for satisfactory substitution is that T_s is much smaller than the relevant time constant Z_1C_2.

$$T_s \ll Z_1C_2, \quad \text{usually} \quad T_s \leqslant 0.1Z_1C_2 \qquad \text{(A.8a)}$$

Since $T_s = Z_s (C_sl_s) = Z_sC_2$, it follows that

$$Z_s \leqslant 0.1Z_1 \qquad \text{(A.8b)}$$

The substitution line is then defined by Z_s and T_s.

A2.3   For a shunt-connected inductance (L_2), the stub line is a short-circuited line satisfying the conditions

$$Z_s = L_2/T_s \geqslant 10Z_1 \qquad \text{(A.9)}$$

A3. REPRESENTATION OF PROTECTIVE DEVICES IN LATTICE DIAGRAMS

The operation of non-linear and voltage-sensitive impedance discontinuities, such as surge diverters or sparkgaps, can be handled in lattice diagrams by the method of voltage cancellation waves.

The sparkgap is a switch, open for voltages less than its sparkover voltage and closed when this voltage is exceeded. Switch closure is represented by the injection of a voltage cancellation wave whose magnitude is determined by the negative value of the difference between the voltage which would appear across the switch if it were open and the voltage developed across the impedance of the protective device after the switch has closed. Again, an appropriate Thevenin-circuit aids the analytical work. An adaptation of *Figure A.1* leads to *Figure A.2*: the Thevenin-voltage is the open-switch voltage $V_{\text{o.c.}}$, the Thevenin-impedance is the impedance of the system viewed from the

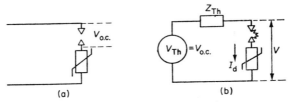

Figure A.2. Surge diverter (a) and equivalent Thevenin circuit (b)

terminals of the protective device, and the impedance of the latter is the load impedance of the circuit. At each instant, $V_{o.c.}$ is compared with the sparkover voltage, and if it is exceeded, the Thevenin-circuit is solved to find the voltage V across the load impedance; the cancellation wave is $(V - V_{o.c.})$. This wave is then impressed on all line elements connected to the same node as the sparkgap. The application of this technique can be seen in *Figure 7.6*.

A4.   TRAVELLING WAVES ON MUTUALLY COUPLED PARALLEL CIRCUITS

A4.1   The travelling wave equations for two conductor ground loops are:

$$v_1 = Z_{11}i_1 + Z_{12}i_2 \qquad (A.10a)$$

$$v_2 = Z_{21}i_1 + Z_{22}i_2 \qquad (A.10b)$$

The following equations A.11 for self and mutual surge impedance are based on perfectly conducting ground but could readily be adjusted

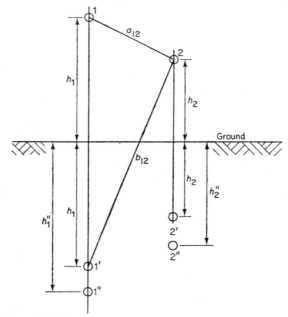

Figure A.3. Calculation of self and mutual surge impedances:
1, 2, conductors; 1′, 2′, image conductors for perfectly conducting ground;
1″, 2″, image conductors for real ground

to reality by setting the image conductors in *Figure A.3* at the correct (lower) depth.

$$Z_{11} = 138 \log 2h_1/r_1 \tag{A.11a}$$

$$Z_{12} = Z_{21} = 138 \log b_{12}/a_{12} \tag{A.11b}$$

$$Z_{22} = 138 \log 2h_2/r_2 \tag{A.11c}$$

A4.2 The surge impedance of two ground wires can be derived from equations A.10 by assuming that, for strokes to tower, they share the current equally; then

$$v = \tfrac{1}{2}(v_1 + v_2) = \tfrac{1}{4}(Z_{11} + 2Z_{12} + Z_{22})i$$

and the self surge impedance of the two wires in parallel, by the use of equations A.11:

$$Z_{eq} = 138 \log (4b^2 h_1 h_2/r_1 r_2 a^2)^{1/4} \tag{A.12}$$

For $h_1 = h_2$ and $r_1 = r_2$

$$Z_{eq} = 138 \log (2bh)^{1/2}/(ar)^{1/2} \tag{A.13}$$

Comparison of equation A.13 with equation A.11a allows the interpretation that the two parallel conductors can be replaced by an equivalent conductor of height $(bh/2)^{1/2}$ and equivalent radius $(ar)^{1/2}$. The latter is clearly much greater than the real one.

Example For one ground wire with average height $h = 30.5$ m and $r = 0.63$ cm, $Z_{11} = 550 \,\Omega$. For two ground wires at the same height, at a distance of $a = 9.15$ m, $Z_{eq} = 332 \,\Omega$, i.e. greater than $\tfrac{1}{2}Z_{11}$. In comparison, a phase conductor with $h = 21.4$ m and $r = 1.5$ cm, would have $Z_{11} = 475 \,\Omega$.

A4.3 If a conductor *1* is struck by lightning, current will flow in both directions. Any parallel conductor *2* will have a voltage v_2 induced on it; if ungrounded at that point, $i_2 = 0$ in equations A.10 and

$$v_1 = Z_{11}i_1; \quad v_2 = Z_{12}i_1$$

hence

$$v_2 = (Z_{12}/Z_{11})v_1 = Kv_1 \tag{A.14}$$

K is termed the 'coupling factor'.

Similarly the coupling factor between two ground wires *1* and *2* and a conductor *3* is calculated from

$$v_3 = Z_{13}i_1 + Z_{23}i_2 + Z_{33}i_3$$

By introducing $i_1 = i_2 = \frac{1}{2}i$ and $i_3 = 0$ the equivalent mutual surge impedance can be found,

$$Z_{m\,eq} = 138 \log (b_{13}b_{23}/a_{13}a_{23})^{1/2} \qquad (A.15)$$

The a's are the distances between the subscripted conductors and the b's the distances between the ground wires and the image of conductor 3. The coupling factor is then the ratio of $Z_{m\,eq}$ to Z_{eq} from equation A.13.

Appendix B

DATA FOR 220 kV TRANSMISSION LINE USED IN LIGHT-
NING PERFORMANCE CALCULATIONS

In Examples 4.1, 5.1–5.5.
Tower configuration: see *Figure B.1*
Ground wire: one, sag 7.6 m (25 ft)
Conductors: steel cored aluminium, sag 9.15 m (30 ft)

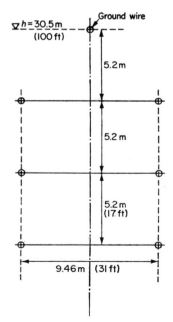

Figure B.1. 220 kV transmission line, tower top configuration

Average span length: 366 m (1200 ft)
Insulators: 15 discs, 254×127 mm (10×5 in)
Thunderday level: 27
Tower footing resistance distribution (power frequency values):

Ohms	0–10	10–20	20–30	30–40	40–50	50–60	60–70	70–80	81+
Per cent frequency	28	16	16	11	10	10	4	3	2

Observed outage rate: 1.02/100 km years (based on 1500 km years of operating experience)
Double-circuit outages: 33% of total

Index

Arc quenching
    by arc suppression coil, 47
    in surge diverter, 101
    by wood insulation, 81
Attenuation of lightning surges, 63–65

Backflashover, 48, 72, 76

Cable-connected equipment, 115
Capacitance
    of insulators, 108
    of transformers, 107
Charge
    above tower, 55
    bound, 49
    trapped, 19–22
Cloudflash, 5
Confidence limits, 31
Contamination of insulation, 34, 103
Corona, 4, 58, 63–65
Counterpoise, 56, 76
Coupling factor, 57, 58, 123

Discharge,
    disruptive, 27, 28
    voltage (of surge diverter), 99
Distance effect in stations, 106, 108
Distortion of travelling waves, 65–68
Double-circuit lines, 79

Earth fault factor, 13

Examples,
    AIEE method, 75
    outage rate, 82
    Monte Carlo method, 80
    shielding, 54, 58
    station co-ordination, 110–112
    switching surge flashover rate, 87
    tower insulation design, 88–90
    unshielded line, 58, 71
    voltage stressing insulation, 58
    insulation strength of wood, 79

Ferranti effect, 12
Flashover voltage
    ambient effects, 34
    critical, 28
    test values, 35–42
Frequency rise, 10

Ground clearance, 91, 93
Groundfault, 13
Groundflash density, 8

Impulse
    insulation level, 43–45, 105
    withstand test, 32, 42
Insulation,
    non-self-restoring, 42
    self-restoring, 28
    standards, 44–45
    test values, 35–45

Lattice diagram, 59, 63, 107, 121

Leader, stepped, 4
Lightning arrester, *see* surge diverter
Lightning characteristics, 5–8
Lightning phenomena, 4, 5
Lightning stroke
    counter, 8
    to ground, 49
    to ground wire, 60, 61
    to phase conductor, 47
    to tower, 55
Load rejection, 10–11

Metifor, 77
Monte Carlo method, 76–77
Model studies, 62

Outage rate, 71
    acceptable, 80
    sustained, 80
Overvoltage,
    energising, 18–20
    factor, 15
    fault initiation, 92
    harmonic, 15
    induced, 49
    lightning, 4
    protection, 97
    switching, 16, 17
    temporary, 9–15

Pollution, *see* contamination
Prediction methods,
    AIEE Committee, 72–74
    Anderson *et al.*, 77
    Clayton and Young, 76
    Monte Carlo 76, 79, 80
Pre-insertion resistor, 21, 22
Probability
    density function, 31, 32
    of flashover, 28
    normal distribution, 28, 29
    reference, 30
    withstand, 32
Power follow current, 99, 101
Protective device, 95–97
Protective level, 95, 99, 105, 109
Protected zone, 104

Reclosing automatic, 2, 80
Rod gap, 97, 98, 114

Saturation effect in iron, 14, 15
Saturation of switching surge strength,
    2, 40, 84
Self-excitation of synchronous genera-
    tor, 13
Self-protecting station, 114, 115
Series capacitor, 11, 12
Shielding, 50–54
Shunt reactor, 11, 12, 23
Standard deviation, 28, 30
Standard impulse, 36, 39
Station
    clearances, 113, 114
    entrance protection, 114
    self-protecting, 114–115
    shielding, 104
Striking distance, 4, 5, 51, 52
Surge diverter, 98–103
    active gap, 101, 102
    tests, 102, 103
Surge impedance, 119, 123
Switching surge control, 21–24

Thevenin theorem, 48, 99, 100, 120,
    122
Thunderday level, 7, 8
Tower
    footing resistance, 56–60
    surge impedance, 56
    window insulation, 88–90
Transient network analyser, 62
Travelling wave theory, 119–124

Voltage
    conventional withstand, 27
    extra high, 9
    statistical withstand, 30
    ultra-high, 92, 93
Voltage-time characteristic, 96–97

Wood pole line, 77, 81